Properties of Double Stars

Per Aspera ad Astra

Plate I. Tailpiece of the 24-inch refractor at Sproul Observatory used for photographic parallax determination. Notice the plate holder and the guiding eyepiece. (Courtesy Dr. P. van de Kamp.)

Properties of Double Stars

A SURVEY OF PARALLAXES AND ORBITS

LEENDERT BINNENDIJK

Professor of Astronomy

University of Pennsylvania

Philadelphia

UNIVERSITY OF PENNSYLVANIA PRESS

Published in Great Britain, India, and Pakistan
by the Oxford University Press
London, Bombay, and Karachi

Library of Congress Catalogue Card Number: 58-8011

Printed in Great Britain
by W. & J. Mackay & Co Ltd, Chatham

Preface

THIS BOOK IS BASED ON LECTURES GIVEN BY THE AUTHOR IN an advanced year course for students who had finished at least a descriptive course in astronomy and who had the necessary basic knowledge in physics and mathematics.

The intention has been to give the student an understanding of the double star problem, beginning with the precautions one has to take even before observations are started and concluding with the final results of the orbital elements. As a consequence, considerable space is used to explain provisional solutions. The definitive solution is then described using the least squares method of the professional astronomer.

In Chapters I, III, and V, the student will find an introduction to astrometry, spectroscopy, and photometry, respectively, as a preparation for the observational techniques and the reductions to be carried out. In the same chapters a rather complete summary of methods of parallax determination is given, because it is important to know the distance both for a binary and for a single star. Some other topics are only touched upon. These chapters can be read in this succession if desired.

In Chapters II, IV, and VI, the different methods of orbital determination which are still in use are discussed and fundamental properties, like mass, size, and density, are studied.

The standard nomenclature has been followed as far as possible. After each chapter selected references are included. No completeness is intended here. As a rule the original publications are mentioned and those which give a summary of the subject or have an extensive bibliography.

It is a privilege to extend my sincere thanks to three astronomers who are experts in the three fields of double stars covered in this volume and who have given me very valuable assistance in the preparation of the manuscript. They are Dr. Peter van de Kamp of Swarthmore College for Chapters I and II, Dr. Dean B. McLaughlin of the University of Michigan for Chapters III and IV, Dr. F. Bradshaw Wood of the University of Pennsylvania for Chapters V and VI. It is a pleasure to express my gratitude to Dr. William Blitzstein, Dr. Robert H. Koch and Mrs. Beverly B. Bookmyer, all of the University of Pennsylvania, for their help during the reading of the proofs.

L. BINNENDIJK

University of Pennsylvania
Philadelphia

Contents

Plates

Abbreviations

A.A.A.S.	*American Association for the Advancement of Science*
A.J.	*Astronomical Journal*
A.N.	*Astronomische Nachrichten*
Ap. J.	*Astrophysical Journal*
B.A.N.	*Bulletin of the Astronomical Institutes of the Netherlands*
I.A.U.	*International Astronomical Union*
L.O.B.	*Lick Observatory Bulletin*
M.N.	*Monthly Notices of the Royal Astronomical Society*
Mt. W.	*Mount Wilson Observatory*
P.A.S.P.	*Publications of the Royal Astronomical Society of the Pacific*
Pop. Astr.	*Popular Astronomy*
R.A.S.C.	*Royal Astronomical Society of Canada*
Z. f. Aph.	*Zeitschrift für Astrophysik*

Properties of Double Stars

I

Astrometry

ASTROMETRY MEANS POSITION DETERMINATION FOR THE purpose of deriving the proper motion, and parallax of a star, and in addition the orbital motion for a double star. J. Bradley's effort to obtain a measurable parallax led to the discovery of aberration and nutation, both of which are much larger shifts in stellar positions than the annual parallax. W. Herschel's attempt to measure parallax led to the discovery of the physical double stars. In 1838 the first parallaxes were measured with a meridian circle and with a heliometer. Those were thus visual observations. Now we observe the parallax only by photographic means. However, in case of the orbital motion of double stars, both the visual and photographic methods are used.

1. *Three fundamental formulae of spherical trigonometry.* Since astrometric measures are essentially measures of a star's position on the celestial sphere, we must consider first some of the fundamentals of spherical trigonometry. A spherical triangle is a part of the surface of a sphere bounded by three great circles. Both the angles and the sides of the spherical triangle are expressed in degrees. The derivation of the three fundamental formulae of spherical trigonometry follows:

(1) In Figure 1 the center of the sphere is at O. The spherical triangle ABC has the sides a, b, c. The plane ADE is tangent

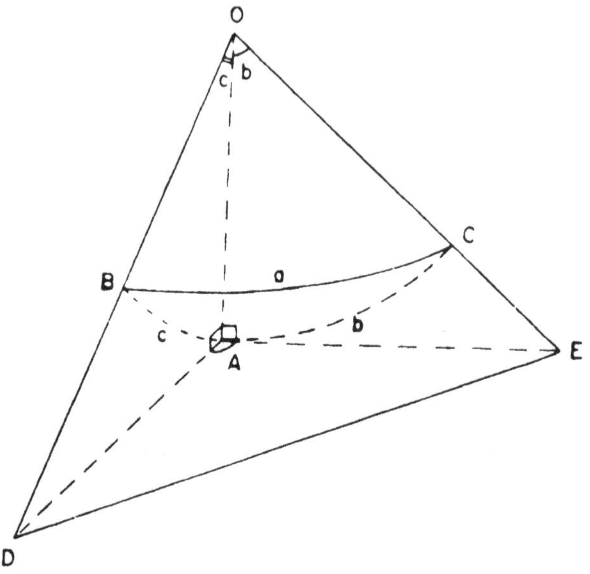

Figure 1. The spherical triangle ABC. The center of the sphere is at O. The plane ADE is tangent to the sphere at A.

to the sphere at A, and therefore ∠ OAE = ∠ OAD = 90°. In Δ ADE and Δ ODE we express DE with the cosine rule (OA=1):

$$DE^2 = \tan^2 b + \tan^2 c - 2 \tan b \tan c \cos A$$
$$DE^2 = \sec^2 b + \sec^2 c - 2 \sec b \sec c \cos a$$
$$= 1 + \tan^2 b + 1 + \tan^2 c - 2 \sec b \sec c \cos a$$

Thus: $-\tan b \tan c \cos A = 1 - \sec b \sec c \cos a$
$$-\sin b \sin c \cos A = \cos b \cos c - \cos a$$

The result is the *cosine rule*, which we will write for all the three sides.

$$\left.\begin{aligned} \cos a &= \cos b \cos c + \sin b \sin c \cos A \\ \cos b &= \cos c \cos a + \sin c \sin a \cos B \\ \cos c &= \cos a \cos b + \sin a \sin b \cos C \end{aligned}\right\} \quad (1)$$

When $C = 90°$ we get:

$$\cos c = \cos a \cos b \quad (2)$$

(2) Square $\sin b \sin c \cos A = \cos a - \cos b \cos c$

$$\sin^2 b \sin^2 c \cos^2 A = \cos^2 a - 2 \cos a \cos b \cos c + \cos^2 b \cos^2 c$$

The left side of the equation can be written as:

$$\begin{aligned} &\sin^2 b \sin^2 c\,(1 - \sin^2 A) = \sin^2 b \sin^2 c - \sin^2 b \sin^2 c \sin^2 A \\ &= (1 - \cos^2 b)(1 - \cos^2 c) - \sin^2 b \sin^2 c \sin^2 A \\ &= 1 - \cos^2 b - \cos^2 c + \cos^2 b \cos^2 c - \sin^2 b \sin^2 c \sin^2 A \end{aligned}$$

Thus we find now:

$$\sin^2 b \sin^2 c \sin^2 A = 1 - \cos^2 a - \cos^2 b - \cos^2 c + 2 \cos a \cos b \cos c$$

This is positive. We define now a positive X so that:

$$X^2 \sin^2 a \sin^2 b \sin^2 c = 1 - \cos^2 a - \cos^2 b - \cos^2 c + 2 \cos a \cos b \cos c$$

Thus: $\dfrac{\sin^2 A}{X^2 \sin^2 a} = 1, \qquad \dfrac{\sin^2 A}{\sin^2 a} = X^2, \qquad + \dfrac{\sin A}{\sin a} = X$

There is only the positive sign because A and a are both $\leqslant 180°$. We find in this way the *sine rule*.

$$\frac{\sin A}{\sin a} = \frac{\sin B}{\sin b} = \frac{\sin C}{\sin c}, \qquad \text{or written out:}$$

$$\left.\begin{aligned} \sin A \sin b &= \sin B \sin a \\ \sin A \sin c &= \sin C \sin a \\ \sin B \sin c &= \sin C \sin b \end{aligned}\right\} \qquad (3)$$

For $C = 90°$ the second equation of the above now gives:

$$\sin A = \frac{\sin a}{\sin c} \qquad (4)$$

(3) We start with the second expression of the cosine rule and use this rule once more.

$$\begin{aligned} &\sin a \sin c \cos B = \cos b - \cos a \cos c \\ &= \cos b - (\cos b \cos c + \sin b \sin c \cos A) \cos c \\ &= \cos b - \cos b \cos^2 c - \sin b \sin c \cos c \cos A \\ &= \cos b \sin^2 c - \sin b \sin c \cos c \cos A \end{aligned}$$

Division by sin c gives the *third rule*:

$$\left.\begin{aligned} \sin a \cos B &= \cos b \sin c - \sin b \cos c \cos A \\ \sin b \cos C &= \cos c \sin a - \sin c \cos a \cos B \\ \sin c \cos A &= \cos a \sin b - \sin a \cos b \cos C \end{aligned}\right\} \qquad (5)$$

For $C = 90°$ the last line gives: $\sin c \cos A = \cos a \sin b$
The cosine rule gave: $\cos c = \cos a \cos b$
Division gives: $\tan c \cos A = \tan b$

$$\text{Thus: } \cos A = \frac{\tan b}{\tan c} \qquad (6)$$

There is also another proof of the third rule. In Figure 2 express x in Δ DBC and Δ DAC with the cosine rule.

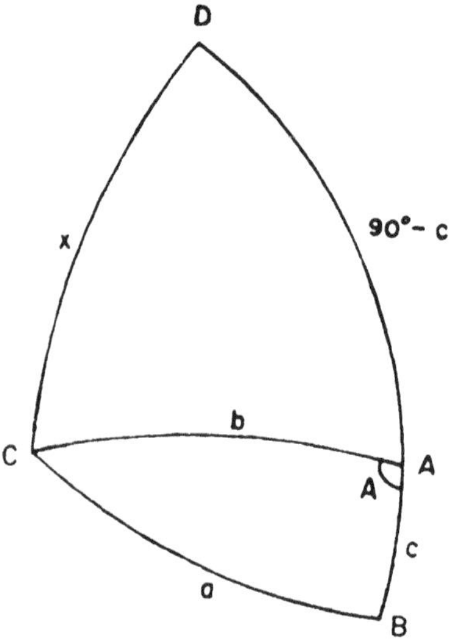

Figure 2. Spherical triangles DBC and DAC.

$$\begin{aligned} \cos x &= \cos a \cos 90 + \sin a \sin 90 \cos B \\ &= \cos b \cos (90-c) + \sin b \sin (90-c) \cos (180-A) \end{aligned}$$

$$\sin a \cos B = \cos b \sin c - \sin b \cos c \cos A$$

2. *Trigonometric parallax.* Stellar parallax means the maximum difference in the lines of sight to a star as seen from the earth and the sun. Let us assume first a circular revolution of the earth around the sun in the plane of the ecliptic. Because the earth's orbit is an ellipse, we thus introduce a very small error, but we will correct for this later. The parallax is also the maximum angle an observer on the star "sees" the distance sun-earth, or the angle the astronomical unit subtends at the star (Figure 3).

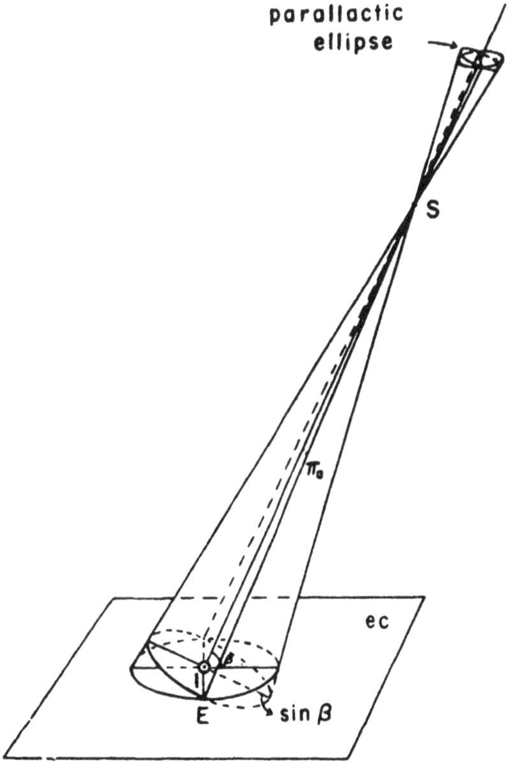

Figure 3. The circular revolution of the earth E in the plane of the ecliptic causes the star to describe a parallactic ellipse in the plane of the sky.

The star as observed from the earth has an apparent movement about the position as observed from the sun which is called the heliocentric position. In space in the direction of the star this will be the exact reflection of the earth's orbit and is thus a circle in a plane parallel to the ecliptic plane. Observed in the plane of the sky we see it thus projected as an ellipse with the semi-major axis equal to the radius of the circle. This semi-major axis is called the parallax. In the line of sight we have also a yearly periodic motion by which the radial velocities are affected. This is independent of the distance and the results are measured directly in kilometers per second. The astronomical unit can be determined from these radial velocity measurements as we will see later.

Let π_a be the absolute parallax. If the parallax is expressed in degrees and the distance r in astronomical units we have according to the definition of parallax (Figure 4):

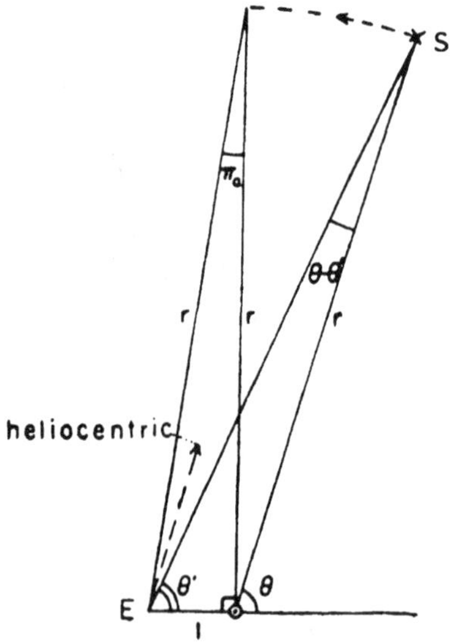

Figure 4. Parallax is the maximum difference in the lines of sights as seen from the earth and the sun.

$$\sin \pi_a = \frac{1}{r} \tag{7}$$

Because π_a is very small we can omit the sine if π_a is expressed in radians. If we express π_a in seconds of arc we have:

$$\pi_a = \frac{206,265}{r}$$

We often use the parsec as a unit of distance. One parsec is the distance at which a star has a parallax of one second of arc. It thus equals 206,265 astronomical units. If π_a is expressed in seconds of arc and r in parsecs the relation becomes simply:

$$\pi_a = \frac{1}{r} \tag{8}$$

In Δ S⊙E we have according to the sine rule:

$$\sin(\theta - \theta') = \frac{1}{r} \sin \theta$$

Again $(\theta - \theta')$ is small. Substitution of (8) gives a relation which is independent of the units used.

$$(\theta - \theta') = \pi_a \sin \theta \tag{9}$$

As seen from the earth the heliocentric direction, the geocentric direction, and the direction towards the sun are in one and the same plane through the earth, sun and star. The geocentric position always will lie on a great circle between the heliocentric position and the sun.

3. *Parallax in celestial longitude and latitude.* Let S (λ, β) be the heliocentric position, S′ (λ', β') the geocentric position, and the sun (⊙, 0) the sun's position in Figure 5. Let P_{ec} be the pole of the ecliptic. SS′⊙ is a great circle and SS′ $= \theta - \theta'$. Further SU is made parallel to the ecliptic. If $\varphi = \angle$USS′ in the little plane triangle USS′ it can be seen that:

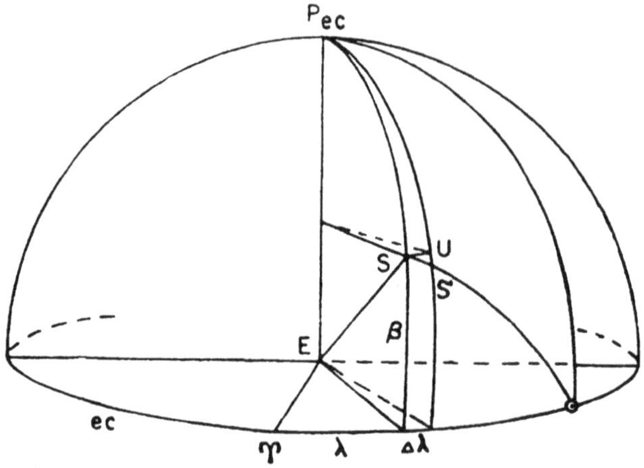

Figure 5. Parallax in celestial longitude and latitude.

$$\left.\begin{aligned} US &= \Delta\lambda \cos\beta = SS' \cos\varphi = \pi_a \sin\theta \cos\varphi \\ US' &= -\ \Delta\beta \quad = SS' \sin\varphi = \pi_a \sin\theta \sin\varphi \end{aligned}\right\}$$

In $\Delta\, SP_{ec} \odot$ according to the sine rule and the third rule we have:

$$\left.\begin{aligned} \sin\theta \cos\varphi &= \sin(\odot - \lambda) \\ \sin\theta \sin\varphi &= \cos(\odot - \lambda) \sin\beta \end{aligned}\right\}$$

Thus we find now, introducing x and y:

$$\left.\begin{aligned} x &= \Delta\lambda \cos\beta = +\pi_a \sin(\odot - \lambda) \\ y &= \Delta\beta \qquad = -\pi_a \cos(\odot - \lambda) \sin\beta \end{aligned}\right\} \tag{10}$$

By squaring these expressions and eliminating the $(\odot - \lambda)$ we find:

$$\frac{x^2}{\pi_a^2} + \frac{y^2}{\pi_a^2 \sin^2\beta} = 1 \tag{11}$$

This is the equation of the parallactic ellipse. The semi-major axis π_a is found to be parallel to the ecliptic; the semi-minor axis $(\pi_a \sin\beta)$, perpendicular to the ecliptic. At the pole of the ecliptic $\sin\beta = 1$ and we get a circle. The geocentric position of the star differs by 180° from the position of the earth in its orbit. In the ecliptic plane $\sin\beta = 0$ and we get a straight line.

The place of the star in the sky with respect to the ecliptic determines the shape of the parallactic ellipse; the distance of the star determines its size. All stellar parallaxes are found to be smaller than 0".76.

4. *Comparison with aberration.* In a similar way we can find the expressions for the aberration:

$$\left.\begin{aligned} x &= \Delta\lambda \cos\beta = -k\cos(\odot - \lambda) \\ y &= \Delta\beta \quad\;\;\; = -k\sin(\odot - \lambda)\sin\beta \end{aligned}\right\} \quad (12)$$

and for the equation of the aberration ellipse:

$$\frac{x^2}{k^2} + \frac{y^2}{k^2 \sin^2\beta} = 1, \qquad k = 20''.5 \quad (13)$$

As far as the aberration is concerned the geocentric position of the star is 90° ahead of the earth in its orbit. Both ellipses have the same shape, but the sizes are very different (Figure 6). The semi-major axis for the aberration ellipse is $k = 20''.5$ depending on the finite velocity of light and on the speed of the earth in its orbit. Stars in a small field have the same aberration ellipse.

Assume $(\odot - \lambda)$ in the first quadrant. The sine and cosine are positive. For parallax we get positive x and negative y, for aberration both x and y are negative and the sine and cosine are reversed here. This means 90° difference in rotation if we omit $\sin\beta$ and look thus at the auxiliary circle. Projection gives the points on the ellipses.

5. *Parallax in right ascension and declination.* We have to find now the expressions using right ascension and declination because these are the coordinates used at the telescope. The right ascension direction at the telescope can be found by stopping the clockdrive during the observation. The star will move then from east to west opposite to the right ascension direction. A bright star will show a trail parallel to the equator at the moment of

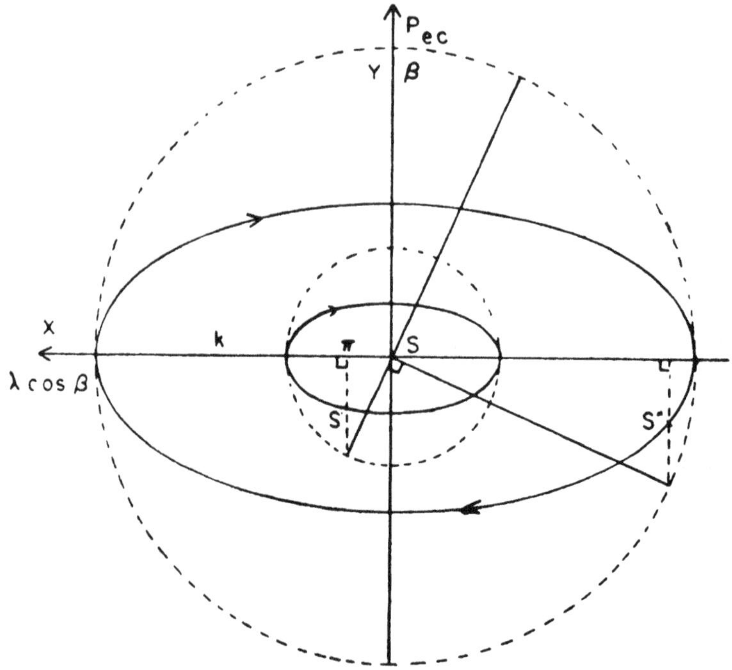

Figure 6. The heliocentric position of the star is S. The geocentric position of the star is S′ on the parallactic ellipse (drawn too large) and S″ on the aberration ellipse.

the photographic exposures. Because essentially we are planning nothing more than a rotation of our axes keeping the same origin, we expect a somewhat more complicated expression.

Let S (α, δ) be the heliocentric position and S′ (α', δ') the geocentric position of the star while (A, D) gives the position of the sun in Figure 7. Further ϵ is the angle between ecliptic and equator. Take US parallel with the equator. If $\psi = \angle$ USS′ then in the small plane triangle USS′ we have:

$$\left.\begin{aligned} \mathrm{US} &= \Delta\alpha \cos\delta = \mathrm{SS'}\cos\psi = \pi_a \sin\theta\cos\psi \\ \mathrm{US'} &= \;\; -\;\Delta\delta \;\;= \mathrm{SS'}\sin\psi = \pi_a \sin\theta \sin\psi \end{aligned}\right\}$$

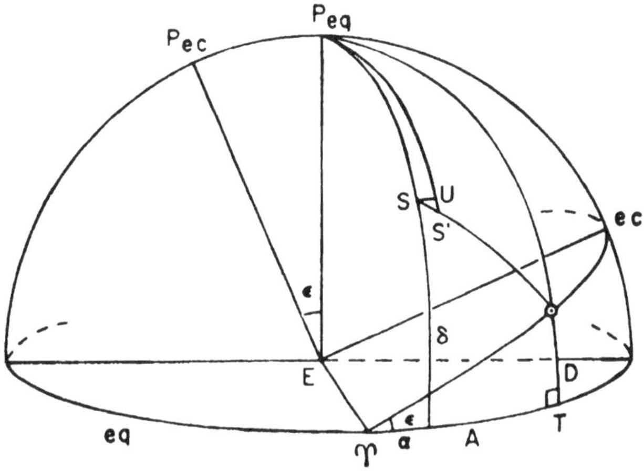

Figure 7. Parallax in right ascension and declination.

In $\Delta\, SP_{eq}\odot$ according to the sine rule and the third rule:

$$\left.\begin{aligned} \sin\theta\cos\psi &= \cos D\sin(A-\alpha) \\ -\sin\theta\sin\psi &= \sin D\cos\delta - \cos D\sin\delta\cos(A-\alpha) \end{aligned}\right\}$$

In $\Delta \Upsilon \odot T$ we have according to the three rules:

$$\left.\begin{aligned} \cos\odot &= \cos D\cos A \\ \sin\odot\sin\epsilon &= \sin D \\ \sin\odot\cos\epsilon &= \cos D\sin A \end{aligned}\right\}$$

Thus we have now:

$$\left.\begin{aligned} \Delta\alpha\cos\delta &= \pi_a\cos D(\sin A\cos\alpha - \cos A\sin\alpha) \\ \Delta\delta &= \pi_a\sin D\cos\delta - \pi_a\cos D\sin\delta(\cos A\cos\alpha \\ &\qquad + \sin A\sin\alpha) \end{aligned}\right\}$$

which give the final expressions:

$$\left.\begin{aligned} \Delta\alpha\cos\delta &= \pi_a\{(\cos\epsilon\cos\alpha)\sin\odot - \sin\alpha\cos\odot\} \\ \Delta\delta &= \pi_a\{(\underline{\sin\epsilon\cos\delta} - \underline{\cos\epsilon}\sin\alpha\sin\delta)\sin\odot \\ &\qquad - \cos\alpha\sin\delta\cos\odot\} \end{aligned}\right\} \quad (14)$$

If we work out our result for the longitude and latitude we can see that the underlined terms are introduced by the rotation.

The ϵ is a constant for a long time interval. The α, δ are constants for the star after correction for precession. The longitude of the sun $\odot$ varies with time and has a period of a year. For a certain date this is also a constant and we find in this way the coordinates of the unit parallactic ellipse $\pi_1 = 1$.

6. *Parallax factors.* These are the coordinates of the unit parallactic ellipse for a certain date. In case of longitude and latitude we have:

$$P_\lambda = \sin(\odot - \lambda), \qquad P_\beta = -\cos(\odot - \lambda)\sin\beta \tag{15}$$

The longitude of the sun can be found in the almanac for each date. The values of λ, β can be computed from the known values of α, δ with help of the parallactic triangle $P_{eq}P_{ec}S$ in Figure 8 with two of the three rules.

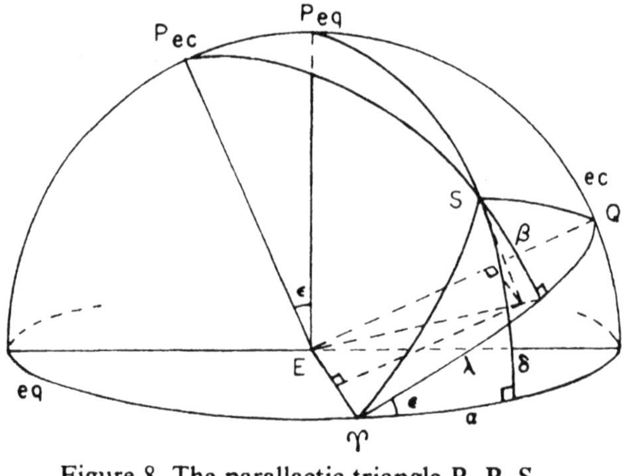

Figure 8. The parallactic triangle $P_{eq}P_{ec}S$.

$$\left.\begin{aligned} \sin\beta &= \cos\epsilon\sin\delta - \sin\epsilon\cos\delta\sin\alpha = \cos SP_{ec} \\ \cos\lambda\cos\beta &= \cos\delta\cos\alpha = \cos S\Upsilon \\ \sin\lambda\cos\beta &= \sin\epsilon\sin\delta + \cos\epsilon\cos\delta\sin\alpha = \cos SQ \end{aligned}\right\}$$

The three expressions are nothing more than the direction cosines of the star with respect to the ecliptic.

In case of right ascension and declination we proceed in the following way. For simplicity define:

$$\left.\begin{array}{ll} p = \cos\epsilon\cos\alpha, & q = -\sin\alpha \\ a = \sin\epsilon\cos\delta - \cos\epsilon\sin\alpha\sin\delta, & b = -\cos\alpha\sin\delta \end{array}\right\} \quad (16)$$

The expressions for the parallax factors are then:

$$\left.\begin{array}{ll} P_\alpha = p\sin\odot + q\cos\odot & \times R \\ P_\delta = a\sin\odot + b\cos\odot & \times R \end{array}\right\} \quad (17)$$

The p, q, a, b are constants for the star under observation. The coordinates P_α, P_δ are thus linear functions of sin ⊙ and cos ⊙ and depend consequently only on the date. The angle ν in Figure 9 is situated between $+ 23\frac{1}{2}°$ and $- 23\frac{1}{2}°$ for the given star. For right ascension 6^h and 18^h this angle is zero because the ecliptic and equator run parallel here.

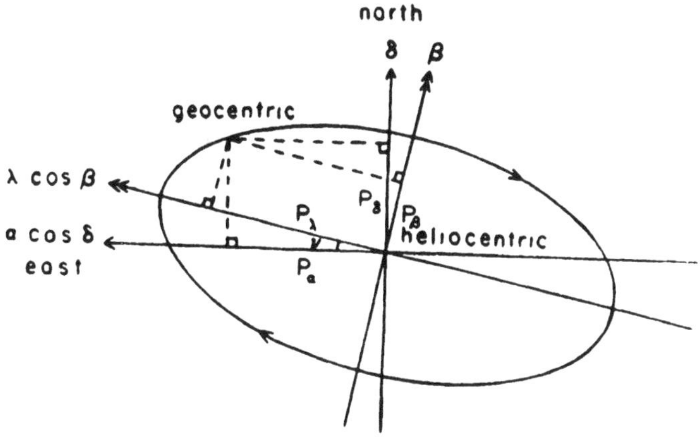

Figure 9. Parallax factors are the coordinates of the unit ellipse.

In our derivation we assumed a circular orbit for the earth with a radius of one astronomical unit. This is not exactly true because we know the true orbit to be an ellipse. The radius vector of the earth's orbit or the distance to the sun R is very close to unity and is given in the almanacs. The largest difference is 0.017 astronomical unit. If we take this into account, we have always $\pi_a R$ instead of π_a in our formulae. It is customary to include the factor R in the final stage in the expression for the parallax factors, which should be computed to three decimals.

One can note here that the parallax will be determined best when P_λ or P_α are near $+1$ or near -1, thus for $\odot - \lambda = 90°$ or 270°, thus $\odot = \lambda + 90°$ or $\odot = \lambda + 270° = \lambda - 90°$. For most accurate results one therefore has to observe the parallax star when the sun differs 90° in longitude from the star. For the star in the meridian this happens at the beginning and end of the night.

7. *Proper motion and parallax.* Let π_r be the relative parallax with respect to the comparison stars. A single star has a linear proper motion μ and a parallactic ellipse with semi-major axis π_r (Figure 10). The heliocentric path of a star is given by $c_x + \mu_x t$, $c_y + \mu_y t$, linear with the time t in years. For the geocentric path we have to add $\pi_r P_\alpha$, $\pi_r P_\delta$. The result is an equidistant spiral. Except for some circumpolar stars we can observe only half of this spiral because the angular distance between the sun and star is too close for the other half of the year.

$$\left.\begin{aligned} \xi &= c_x + \mu_x t + \pi_r P_\alpha \\ \eta &= c_y + \mu_y t + \pi_r P_\delta \end{aligned}\right\} \qquad (18)$$

The known factors are the measured quantities ξ, η, the time t and the parallax factors P_α, P_δ. Unknown are c_x, c_y, μ_x, μ_y, π_r and they have to be determined from the observations. The constants c_x, c_y are added because the proper motion will not go exactly through the origin. A graphical method to determine the constants roughly is as follows.

Take two observations which are exactly one year apart. They give us the proper motion and its components μ_x, μ_y; thus we have now size and direction of the proper motion. Now determine the width of the strip in which all observations are situated. We know also the unit ellipse in its true orientation. Decrease this ellipse in such a way that it fits in the strip. The semi-major axis is the parallax π_r. The straight line of the proper motion gives for $t = 0$ the constants c_x, c_y. In practice we compute this still more accurately. Because the parallax is so small, we have to take many precautions in taking the observations, in the measurement, and in the reduction in order to get reliable results, and these we must study first.

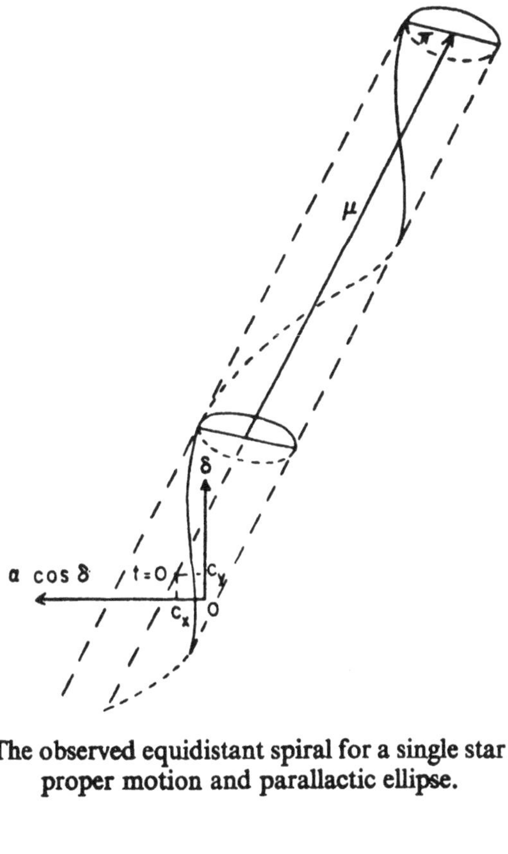

Figure 10. The observed equidistant spiral for a single star consists of proper motion and parallactic ellipse.

8. *Photographic stellar parallaxes.* At present parallaxes of about 6000 stars have been determined. F. Schlesinger was the first to study the basic precautions one has to take in photographic parallax work. One needs a refracting telescope of long focal length because then the scale is large. The scale factor is defined as the number of seconds of arc per mm. It can be found by measuring the distance between two known stars in an open cluster for which the right ascensions differ but the declinations are the same. It is best to take this plate during meridian passage. For accurate work one has to take the refraction into account which in the vertical circle is given by the refraction constant times tan z, where z is the zenith distance and the so-called refraction constant is the atmospheric refraction at $z = 45°$.

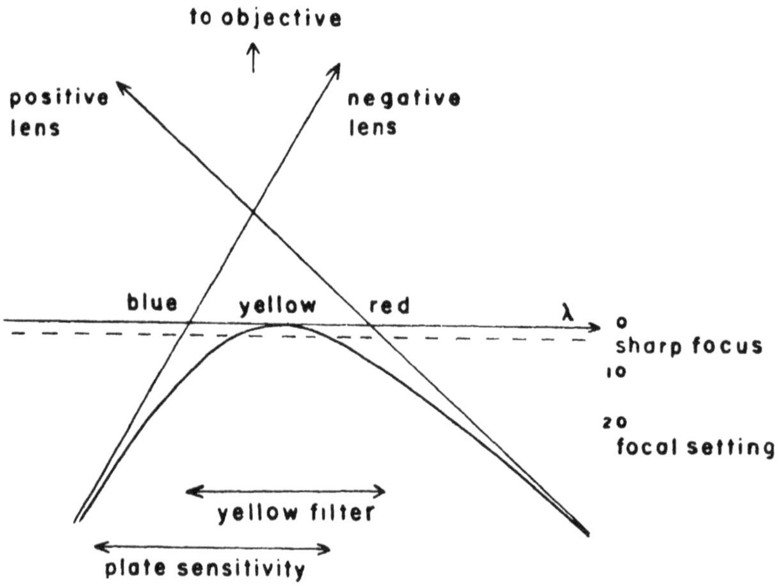

Figure 11. The focal curve of an objective.

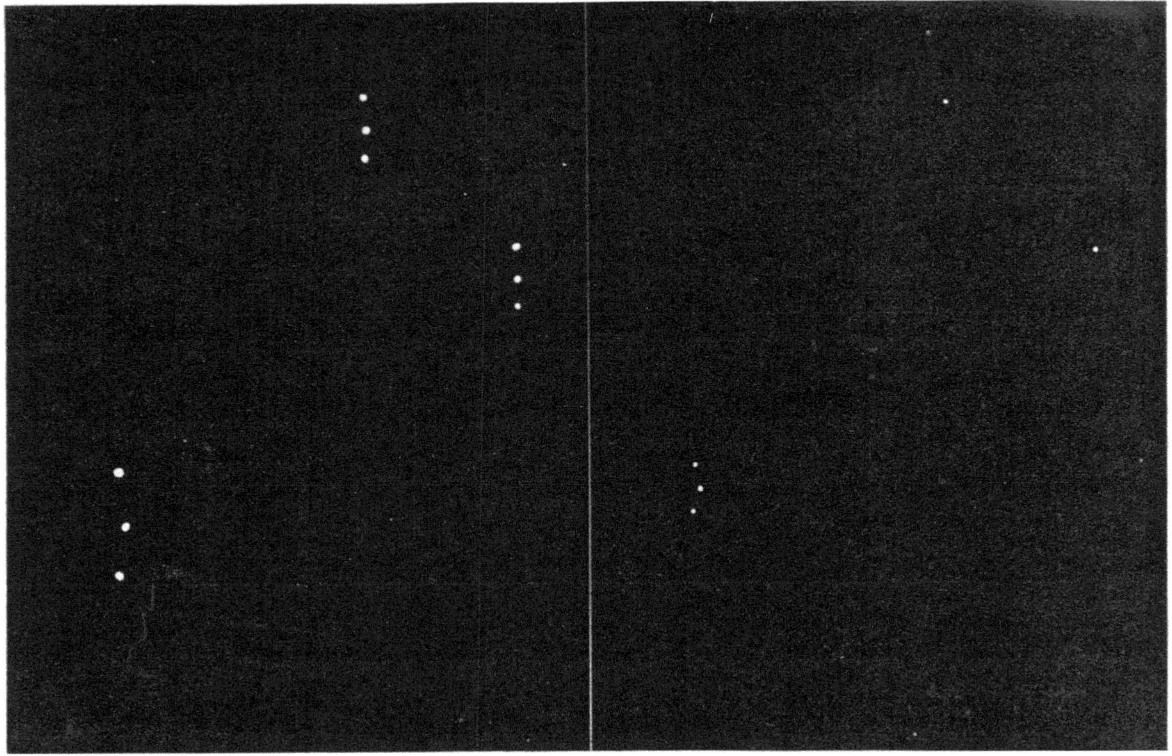

Plate II. Illustration of the parallax and proper motion of Barnard's star. Left: At three epochs in a year the three fields were photographed. Barnard's star (lower left) shows displacement both east-west (parallax) and south-north (proper motion) compared to the two background stars. Right: The same photographs are super-imposed, again illustrating the displacement of Barnard's star in two directions. (Courtesy Dr. P. van de Kamp.)

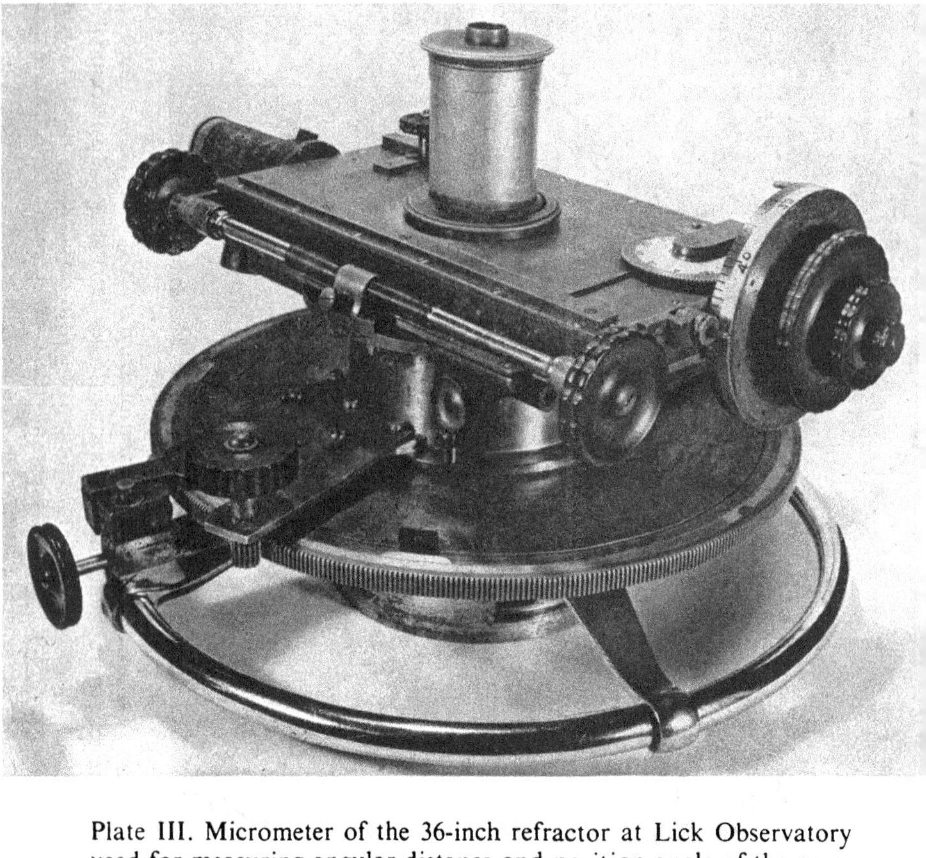

Plate III. Micrometer of the 36-inch refractor at Lick Observatory used for measuring angular distance and position angle of the components of a visual double star. (Courtesy Dr. G. E. Kron.)

Any vertical angular distance suffers a contraction, and any horizontal distance, measured along the great circle suffers a contraction. In the zenith the refraction is zero, but not the differential refraction on the zenith plate. The scale factor in use for various telescopes refers to the zenith. The value 1 mm usually refers to the pitch of the measuring machine.

A photovisual refractor is the best for this kind of work. The focal curve or secondary spectrum of the objective shows the relation between wavelength and focal setting (Figure 11). This can be found by using monochromatic light of a monochromator and observing the focal setting where the image is sharp. A yellow filter absorbs all wavelengths except in the yellow. The plate sensitivity is chosen in such a way that the effective wavelength region which affects the plate is still smaller. We have thus nearly monochromatic light coming in practically the same focus. The images are thus sharp. In this way we overcome the chromatic aberration of the objective. A reflecting telescope does not have this trouble. In this figure we would thus get a horizontal line. However, there we have difficulties with the diffraction images of the secondary mirror system. Therefore, for this work we prefer not to use reflecting telescopes.

The spherical aberration of the objective can be found with help of the so-called Hartmann disk in which holes are made in successive rings. One takes two exposures, far out of focus—one

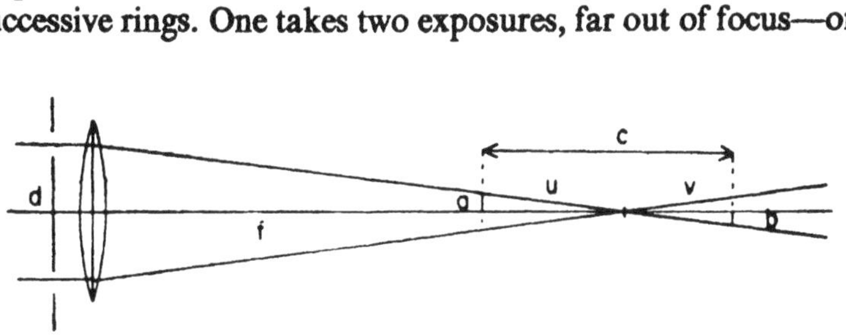

Figure 12. Determination of the spherical aberration of an objective with the Hartmann disk.

on either side of the focal plane (Figure 12). From the pattern of the images on the plate one can find the best focal setting for each of the rings.

$$u : v = a : b, \qquad u : (u + v) = a : (a + b)$$

$$u = \frac{a}{a + b}(u + v) = \frac{ac}{a + b}$$

9. *Observation.* The sky condition has to be judged before we start observing. The seeing must be good in order to get small images on the plate. There are two ways in which we can get large images. The star image may jump around in the field, an effect caused by turbulence in our atmosphere due to a strong wind at high altitude. Or the star image may also look steady but blown up. This latter effect is caused by high humidity in the atmosphere. We have unusually fine seeing when the sky is slightly overcast and the transparency bad.

The field covered by the plate is so small that the refraction, aberration, nutation and precession for all stars are nearly the same. We measure in a relative way and take into account the first order differences. However, the refraction constant depends on the wavelength and our light is not strictly monochromatic. This causes the atmospheric dispersion (d tan z). The star images are thus in reality small spectra in a vertical direction. The red rays of the star show the smallest deviation of the original direction. Therefore the blue rays appear closer to the zenith than do the red rays. For a zenith distance of 60° the distance between the G-line (4308 A) and the D-line (5893 A) is 1″.4.

In $\Delta\, P_{eq}ZS$ of Figure 13 let $x = \angle P_{eq}SZ$. We have according to the sine rule:

$$\sin x = \frac{\sin \tau}{\sin z} \cos\varphi$$

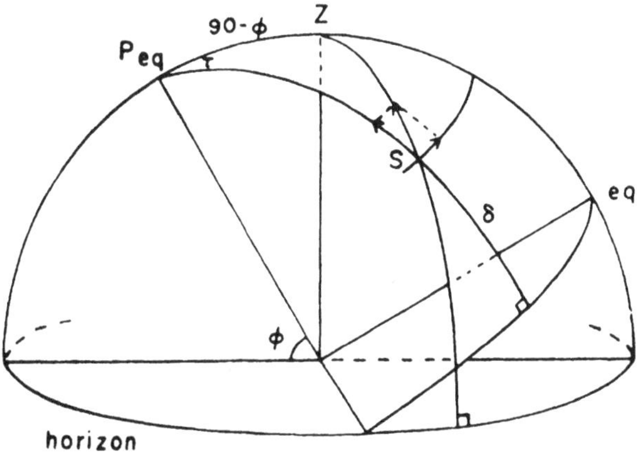

Figure 13. The spherical triangle $P_{eq}ZS$.

We have to project the atmospheric dispersion on the right ascension direction and have thus:

$$d \tan z \sin x = d \sin \tau \cos \varphi \sec z$$
$$\approx \tau d \cos \varphi \sec (\varphi - \delta)$$

The dispersion in right ascension is small but approximately linear with the hour angle. In declination the dispersion is large but practically constant. In relative work this is an advantage, and the declination observations are less affected. We can overcome most of the trouble by the following precautions: (1) Use a yellow filter. (2) Observe not more than one hour from the meridian. (3) Choose the comparison stars of the same color as the parallax star.

The observer has to guide or to keep the star on the crosswires in the field. This is necessary because the man-made clock drive does not work as regularly as the sky rotates. By a mechanical arrangement the plate holder can be moved with respect to the whole telescope so that quick resetting is possible (Plate I).

Guiding errors have a different effect on bright images than on faint images. Because the guiding error lasts only a short time for a good observer, one cannot notice it on the faint images. It is therefore necessary to dim the bright nearby parallax star until it gives the same brightness on the plate as the distant comparison stars. This is done with a rotating sector in front of the plate near the center. One has to take care that the star's image falls under the sector and the images of the comparison stars outside the sector.

There is another reason why the brightness of all stars should be the same. The sensitivity of the filter-plate combination usually falls off more sharply toward longer wavelengths than toward the shorter ones (Figure 14). A faint star will have an effective wavelength around the maximum of the sensitivity curve. The bright star will have a shorter effective wavelength. Atmospheric dispersion thus will move the bright star more towards the zenith.

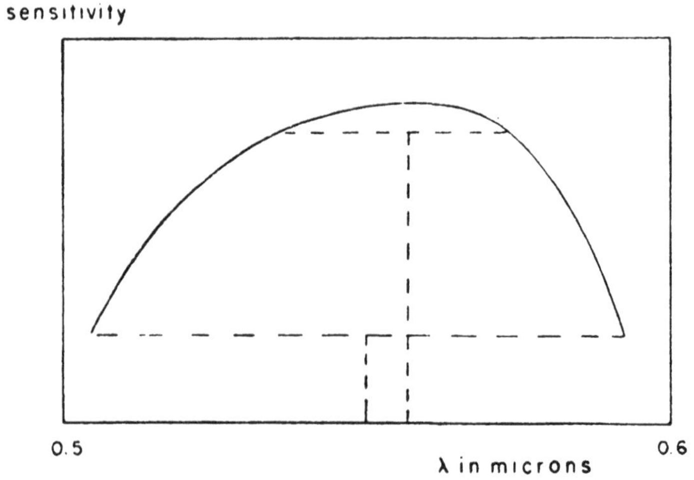

Figure 14. Sensitivity curve of a filter-plate combination used in parallax observation.

After about four exposures in a row, the plate is taken out in a darkroom, rotated 180° in its own plane and again four exposures are taken. This cancels any emulsion shift if this shift is linear.

The development is done in tanks in this succession: development 6 minutes at 68°F = 20°C, rinsing 1 minute, hardening 5 minutes, fixing 30 minutes, rinsing 60 minutes. The water has to flow constantly because the fixing salts are heavier than water. It is important that the drying occurs in a natural way, not in draft or near a stove.

10. *Measurement*. The measuring machine has a long screw. The screw error can be determined by measuring the same distance interval between two stars several times on different parts of the screw. This screw error must be small and has to be checked regularly. The screw has to be lubricated with excellent oil.

One of the plates during the middle of the observing interval is taken with a trail, which gives us the direction of the equator at that moment. We will call this our standard plate.

We rotate our plate to a convenient equinox, $t_0 = 1950$ for example, according to the formula:

$$\Delta\theta = -0\overset{\circ}{.}0057\,(t - t_0)\sin\alpha\sec\delta \qquad (19)$$

A position of a star is now measured by bisecting the image. This can be done within 1% of the size of the image, which may be 0.1 mm. After measuring a sequence of stars the plate is rotated 180° and measured again. We always measure along the screw in the same direction. Thus the screw error is partly balanced out. A weight is given according to the appearance of the image on the plate. By adding all the weights we get the weight of the measurements on the plate.

11. *Dependences*. Dependences are relative weights given according to the positions of the comparison stars with respect to the

position of the parallax star in the configuration on the standard plate. For three comparison stars these dependences have a geometric meaning (Figure 15). We will find that they are the

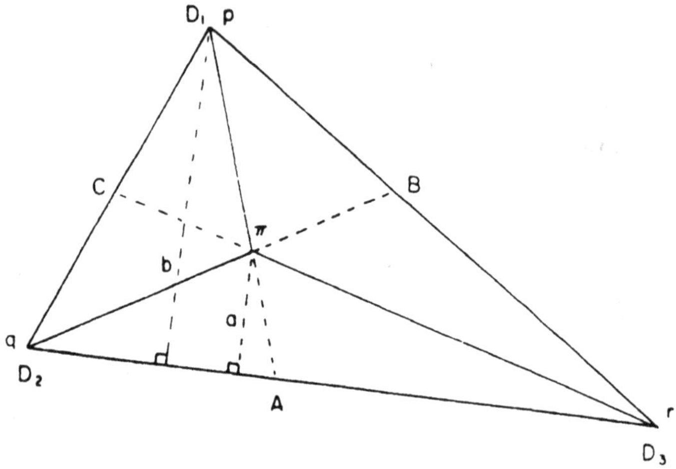

Figure 15. The dependence D is the area of the opposite triangle expressed in the area of the whole triangle as a unit.

areas of the opposite triangles expressed in terms of the whole triangle as a unit. A provisional value of these dependences can then be found from the following ratio of lines:

$$D_1 = \frac{\Delta \pi \mathrm{rq}}{\Delta \mathrm{prq}} = \frac{a}{b} = \frac{\pi \mathrm{A}}{\mathrm{pA}} \tag{20}$$

Similar expressions hold for the other two dependences. The sum of the dependences equals unity.

Let (x_1y_1), (x_2y_2), (x_3y_3) be the coordinates of the three comparison stars measured on the standard plate and X_0Y_0 those of the parallax star (Figure 16). For another plate we will write with primes, thus $(x'_1y'_1)$, $(x'_2y'_2)$, $(x'_3y'_3)$ for the comparison stars and $X'Y'$ for the parallax star, which has moved in the time elapsed (Figure 17). The zero point, scale and orientation measured are close to that of the standard plate, but it is impossible to do this exactly. The scale correction a_x is proportional to x_1, the orientation b_x is proportional to y_1, while the zero point correction is c_x. We have then:

$$\left.\begin{aligned} a_x x_1 + b_x y_1 + c_x &= x_1 - x'_1 \qquad \times D_1 \\ a_x x_2 + b_x y_2 + c_x &= x_2 - x'_2 \qquad \times D_2 \\ a_x x_3 + b_x y_3 + c_x &= x_3 - x'_3 \qquad \times D_3 \end{aligned}\right\} \tag{21}$$

$$a_x X_0 + b_x Y_0 + c_x = X - X' \qquad \times -1 \tag{22}$$

respectively for the comparison stars and the parallax star.

After multiplication and adding we find:

$$\begin{aligned} D_1(x_1 - x'_1) + D_2(x_2 - x'_2) + D_3(x_3 - x'_3) - (X - X') = \\ = a_x(D_1 x_1 + D_2 x_2 + D_3 x_3 - X_0) \\ + b_x(D_1 y_1 + D_2 y_2 + D_3 y_3 - Y_0) \\ + c_x(D_1 + D_2 + D_3 - 1) \end{aligned}$$

The elimination of the plate constants a_x, b_x, c_x is effective if:

$$\left.\begin{aligned} D_1 x_1 + D_2 x_2 + D_3 x_3 &= X_0 \\ D_1 y_1 + D_2 y_2 + D_3 y_3 &= Y_0 \\ D_1 + D_2 + D_3 &= 1 \end{aligned}\right\} \tag{23}$$

The solution can be written in the form:

$$\frac{D_1}{\begin{vmatrix} X_0 & x_2 & x_3 \\ Y_0 & y_2 & y_3 \\ 1 & 1 & 1 \end{vmatrix}} = \frac{1}{\begin{vmatrix} x_1 & x_2 & x_3 \\ y_1 & y_2 & y_3 \\ 1 & 1 & 1 \end{vmatrix}} \tag{24}$$

The determinant in the right denominator represents the area of the whole triangle; the determinant of the left fraction equals the area of the triangle opposite to the star associated with D_1. Dependences are here thus the areas of the opposite triangles expressed in the whole triangle as a unit.

For the parallax star we get now what remains over in the equation.

$$X - X' = D_1(x_1 - x'_1) + D_2(x_2 - x'_2) + D_3(x_3 - x'_3)$$
$$= \sum_{i=1}^{3} Dx_i - \sum_{i=1}^{3} Dx'_i$$

We usually use here the symbol [ ] instead of the Σ sign.

$$\left.\begin{array}{l} X = [Dx] + X' - [Dx'] \\ Y = [Dy] + Y' - [Dy'] \end{array}\right\} \qquad (25)$$

The $[Dx]$ is found on the standard plate, and the $[Dx']$ from the other plate. We can thus correct X' to X in our standard system.

12. *Method of dependences.* For more than three comparison stars we cannot give such a geometrical picture. The number of equations is no longer equal to the dependences to be found. We have to proceed by least squares. For simplification we take the zero point in such a way that the average value of the x and the y are zero for the n comparison stars on the standard plate. Thus $[x] = [y] = 0$. This determines the gravity center G of the

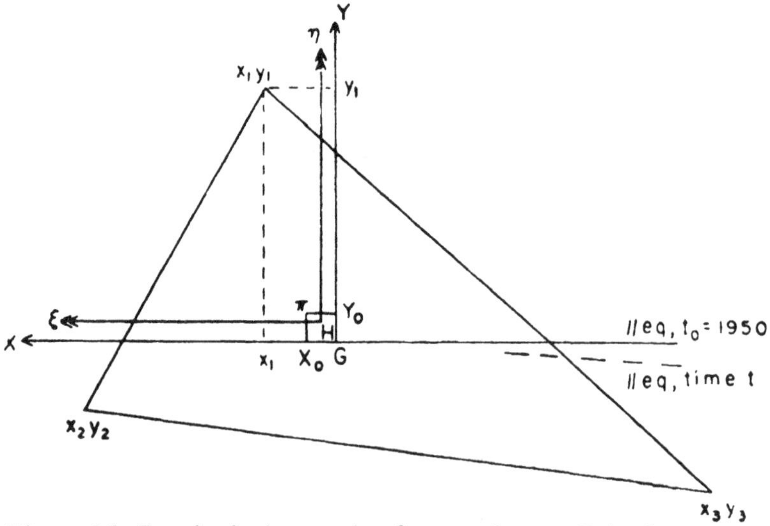

Figure 16. Standard plate and reference frame. G is the gravity center. H is the dependence center.

triangle in Figure 16. We will solve only for the x-coordinate; the y-coordinate goes in the same way. Our linear equation of condition was:

$$\left.\begin{aligned} a_x x + b_x y + c_x &= x - x' \\ a_y x + b_y y + c_y &= y - y' \end{aligned}\right\}$$

For the x-coordinate the normal equations are:

$$\left.\begin{aligned} [x^2]\, a_x + [xy]\, b_x \qquad &= [x(x-x')] \\ [xy]\, a_x + [y^2]\, b_x \qquad &= [y(x-x')] \\ nc_x &= [(x-x')] = -[x'] \end{aligned}\right\}$$

The unknowns a_x and b_x can be solved with help of determinants.

$$a_x = \frac{\begin{vmatrix} [x(x-x')] & [xy] \\ [y(x-x')] & [y^2] \end{vmatrix}}{\begin{vmatrix} [x^2] & [xy] \\ [xy] & [y^2] \end{vmatrix}} \text{ etc.}$$

The $[x^2]$, $[xy]$, $[y^2]$ are constants from measurements on the standard plate. Therefore:

$$a_x = \frac{[y^2]\,[x(x-x')] - [xy]\,[y(x-x')]}{[x^2]\,[y^2] \quad - [xy]^2}$$

$$b_x = \frac{[x^2]\,[y(x-x')] - [xy]\,[x(x-x')]}{[x^2]\,[y^2] \quad - [xy]^2}$$

$$c_x = -\frac{[X']}{n}$$

Another notation is as follows:

$$a_x = \sum_{i=1}^{n} \frac{\{x_i\,[y^2] - y_i\,[xy]\}}{[x^2]\,[y^2] - [xy]^2} (x_i - x'_i)$$

$$b_x = \sum_{i=1}^{n} \frac{\{y_i\,[x^2] - x_i\,[xy]\}}{[x^2]\,[y^2] - [xy]^2} (x_i - x'_i)$$

$$c_x = \sum_{i=1}^{n} \frac{1}{n}(x_i - x_i')$$

Similar expressions are found for a_y, b_y, c_y. The measurements of the comparison stars give us the plate constants a_x, b_x, c_x. We need thus at least three equations, and thus three comparison stars. Let us consider now our parallax star, which moved during the time interval between standard plate and the other plate. Let X, Y be the new position reduced to the standard plate system.

$$\left.\begin{aligned} X &= X' + a_x X_0 + b_x Y_0 + c_x \\ Y &= Y' + a_y X_0 + b_y Y_0 + c_y \end{aligned}\right\} \qquad (26)$$

We could thus compute the a_x, b_x, c_x, etc. and place these in the equations. We could compute our values X, Y on the correct coordinate system of the standard plate. In other words we would find one observed point on the spiral. However, the computation of the a_x, b_x, c_x, etc. would have to be carried out for each plate and this represents a tremendous amount of work. Furthermore we are not fundamentally interested in those plate constants but only in the X, Y. Therefore in practice we proceed as in the next paragraph.

13. *Plate solutions.* Let us substitute the mathematical expressions for the a_x, b_x, c_x into the formula for X. We get then:

$$X = X' + \sum_{i=1}^{n} \left\{ \frac{X_0(x_i[y^2] - y_i[xy]) + Y_0(y_i[x^2] - x_i[xy])}{[x^2]\,[y^2] - [xy]^2} + \frac{1}{n} \right\} \times (x_i - x'_i)$$

We introduce now as dependences $D_1, D_2 \ldots . D_n$.

$$D_i = \frac{X_0(x_i[y^2] - y_i[xy]) + Y_0(y_i[x^2] - x_i[xy])}{[x^2]\,[y^2] - [xy]^2} + \frac{1}{n} \qquad (27)$$

It is obvious that $[D] = 1$ because $[x] = [y] = 0$.

Also $[D^2] =$ minimum because of the least squares procedure. Notice that only standard plate data enter in the formula. The dependence reduction can now be written as:

$$\left.\begin{aligned} X &= X' + [D(x-x')] = [Dx] + X' - [Dx'] = [Dx] + \xi \\ Y &= Y' + [D(y-y')] = [Dy] + Y' - [Dy'] = [Dy] + \eta \end{aligned}\right\} \quad (28)$$

The dependence center H in Figures 16 and 17 has as coordinates $[Dx]$, $[Dy]$. With respect to this center the coordinates become for the observed point on our spiral:

$$\left.\begin{aligned} \xi &= X' - [Dx'] = X' - (D_1x'_1 + D_2x'_2 + D_3x'_3 + \ldots) \\ \eta &= Y' - [Dy'] = Y' - (D_1y'_1 + D_2y'_2 + D_3y'_3 + \ldots) \end{aligned}\right\} \quad (29)$$

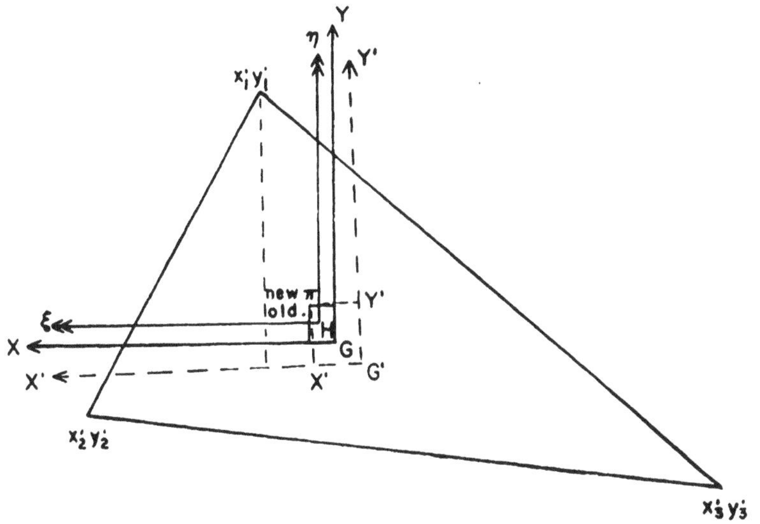

Figure 17. Provisional reference frame of another plate. The parallax star moved from the old position to the new position π.

14. *Computation of dependences.* We use the measurements of the comparison stars and parallax star on the standard plate. Call:

$$f_i = \frac{x_i\,[y^2] \quad - y_i\,[xy]}{[x^2]\,[y^2] - [xy]^2}, \qquad g_i = \frac{y_i\,[x^2] \quad - x_i\,[xy]}{[x^2]\,[y^2] - [xy]^2} \tag{30}$$

These depend on the coordinates of the comparison stars only. We see that $[fx] = [gy] = 1$. The dependences can now be written in the following linear relation:

$$D_i = f_i X_0 + g_i Y_0 + \frac{1}{n} \tag{31}$$

For the three comparison stars in a triangle with equal sides and the parallax star in the gravity center we find $D_i = 1/3$. The f_i, g_i tell us about the asymmetry; the X_0, Y_0, describe the deviation of the parallax star from the gravity center. Because $[x] = [y] = 0$, it must follow strictly: $[D] = 1$. For the standard plate we tried to make $\xi = 0$, $\eta = 0$, so that $X = X_0$, $Y = Y_0$ and within the rounding off errors it follows $[Dx] = X_0$, and $[Dy] = Y_0$.

We can also compute the annual variation ΔD by differentiating:

$$\Delta D_i = f_i\,\mu_x + g_i\,\mu_y \tag{32}$$

with the controls: $[\Delta D] = 0$, $[\Delta Dx] = \mu_x$, $[\Delta Dy] = \mu_y$.
For small proper motions of the parallax star it is not necessary to change the D's, but for large proper motion two or three sets of D's may be advisable. We compute the dependences to three decimals using only the standard plate. For the other plates we then find the position ξ, η with those dependences on the correct coordinate system.

15. *Number of comparison stars.* The solution for every plate had the form:

$$\xi = X' - D_1 x'_1 - D_2 x'_2 - D_3 x'_3 - \ldots - D_n x'_n$$

The square of the probable error of this solution is:

$$\epsilon_{\xi}^2 = \epsilon_{X'}^2 + D_1 \epsilon_{x'_1}^2 + D_2 \epsilon_{x'_2}^2 + D_3 \epsilon_{x'_3}^2 + \ldots + D_n \epsilon_{x_n'}^2$$

Assume that the probable errors of the measurements of all stars are the same namely ϵ; it follows:

$$\epsilon_{\xi}^2 = \epsilon^2(1 + [D^2]), \qquad \epsilon_{\xi} = \epsilon \sqrt{(1 + [D^2])}$$

If the stars are well divided over the plate, so that $f_i = g_i = 0$ we have $D_i = 1/n$, and we get then:

$$\epsilon_{\xi} = \epsilon \sqrt{(1 + \frac{n}{n^2})} = \epsilon \sqrt{(1 + \frac{1}{n})} \tag{33}$$

For three equal dependences the square root is 1.15; for four we find 1.12, etc. Usually one takes three or four comparison stars and measures the parallax star twice. More comparison stars take too much measuring time for the slight increase in accuracy.

How does the distribution of the comparison stars over the plate influence the probable error of the result? We can compute this for different cases. It can be seen then that the distribution and thus the dependences may be quite unequal, and that the parallax star does not have to be situated in the gravity center. Only when the parallax star lies outside the configuration do we get negative dependences which we have to avoid.

16. *Relative and absolute trigonometric parallax.* Usually one takes two glass plates per night. Taking more is useless because of systematic night errors. There are also plate errors and measurement errors. The order of these probable errors are respectively: $\pm$ 0".012, $\pm$ 0".023, $\pm$ 0".015, combined $\pm$ 0.030. All measurements for one night are averaged to a mean night value.

A least squares solution is made for which the equation of condition is according to formula (18):

$$\left.\begin{aligned} \xi &= c_x + \mu_x t + \pi_r P_\alpha \\ \eta &= c_y + \mu_y t + \pi_r P_\delta \end{aligned}\right\}$$

One solves these equations for each coordinate first separately and then combined. The relative parallax π_r, the proper motion (μ_x, μ_y) and the zero point (c_x, c_y) are found in this way. Also the probable errors follow from the computation (Plate II).

We measured the relative ellipse of the parallax star with respect to the comparison stars. Those comparison stars are far away but not at an infinite distance. In other words they have a small absolute parallax, which has to be added to the relative parallax of the π star to reduce this value to an absolute one, π_a. Figure 18 shows why one has to add this correction. It is still better to add the average parallax value of the comparison stars or $[D\pi_c]$. The parallaxes of the comparison stars are found in a statistical way from proper motions and are given as a function of apparent magnitude and galactic latitude. The most recent values are given by A. N. Vyssotsky and E. T. R. Williams derived from secular parallaxes, and by L. Binnendijk from peculiar motions according to a method of J. H. Oort.

The probable error of a good parallax determination is of the order of $\pm$ 0".005. However, it was found that there are systematic differences between results of different observatories for the same star. The amount may be 0.003 and can be explained if the number of evening and morning plates are not the same. There is also a periodic error for which one can apply an empirical correction. Very probably the result is caused by temperature changes in the objectives. Some stars are observed on winter mornings and summer evenings. The reduction curves of different observatories are not the same, which is surprising. These corrections have been applied in the Yale Catalogue of

Parallaxes to produce a uniform system of absolute parallaxes. This catalogue is very important because all other distant measurements are indirect and have to be calibrated with the trigonometric parallaxes. Thus the trigonometric parallaxes form the basis of all distances measured in the universe.

To know those distances in kilometers or miles, the astronomical unit has to be known in kilometers or miles. There are several ways to find this. The orbital motion of the minor planet Eros provides an excellent opportunity to determine this distance. The daily parallax of the sun was found to be 8".80, corresponding to a distance of 149.5 million kilometers.

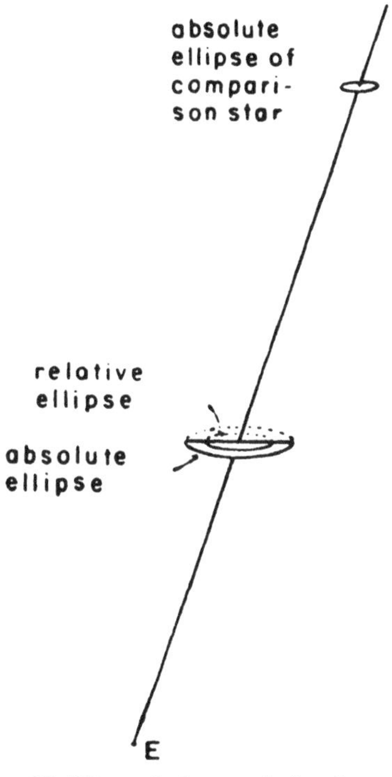

Figure 18. The relative and absolute ellipses.

17. *Proper motions*. The proper motions of many stars have been measured and published in proper motion catalogues. An example is the Radcliffe photographic catalogue. One takes plates at least 20 years apart in time at about the same date of the year. Measures are made relative to the faintest stars on the plate. A correction of the form $(ax + by + c)$ must be applied. The probable error in one coordinate in the Radcliffe catalogue is $\pm$ 0·003; in case of the Pleiades catalogue of proper motions by E. Hertzsprung the probable error is $\pm$ 0".001 in each coordinate.

One next tries to derive absolute proper motions; and a first sequence of plates has been taken at Lick Observatory showing the extra-galactic nebulae. A difficulty which still remains is that the images of stars and nebulae do not look alike.

Accurate proper motions are important for a number of problems. The peculiar motion of a certain star is usually unknown, but for a large number of stars the sum will be zero if the velocity distribution is at random. The problems are thus statistical in nature, and one can find the precession constant, the linear solar motion, the galactic rotation effect, mean parallaxes, etc. A complication appears because the real distribution of stellar velocities shows a preferential motion, which can be explained by the theory of the galactic rotation.

18. *Solar motion and apex*. The point toward which the sun is moving in space with respect to the center of the nearby stars is called the apex of the linear solar motion. It is in the constellation of Herculus. It can be found from proper motions and radial velocities of stars. The stars situated near the apex seem to move towards the sun and will show a negative radial velocity on the average while near the antapex they will be receding and show a positive radial velocity. Exactly 90° away the effect in the radial velocity is zero, while that for the proper motions

is at a maximum, opposite in direction to the solar motion and depending in size on the mean distance of the star group.

In Figure 19 let the sun be at the origin and the rectangular coordinates of a star x, y, z or in polar coordinates r, α, δ. We have then the relations:

$$\left.\begin{aligned} x &= r\cos\delta\cos\alpha \\ y &= r\cos\delta\sin\alpha \\ z &= r\sin\delta \end{aligned}\right\} \tag{34}$$

By differentiation we get:

$$\left.\begin{aligned} \frac{dx}{dt} &= \frac{dr}{dt}\cos\delta\cos\alpha - r\sin\delta\cos\alpha\frac{d\delta}{dt} - r\cos\delta\sin\alpha\frac{d\alpha}{dt} \\ \frac{dy}{dt} &= \frac{dr}{dt}\cos\delta\sin\alpha - r\sin\delta\sin\alpha\frac{d\delta}{dt} + r\cos\delta\cos\alpha\frac{d\alpha}{dt} \\ \frac{dz}{dt} &= \frac{dr}{dt}\sin\delta + r\cos\delta\frac{d\delta}{dt} \end{aligned}\right\}$$

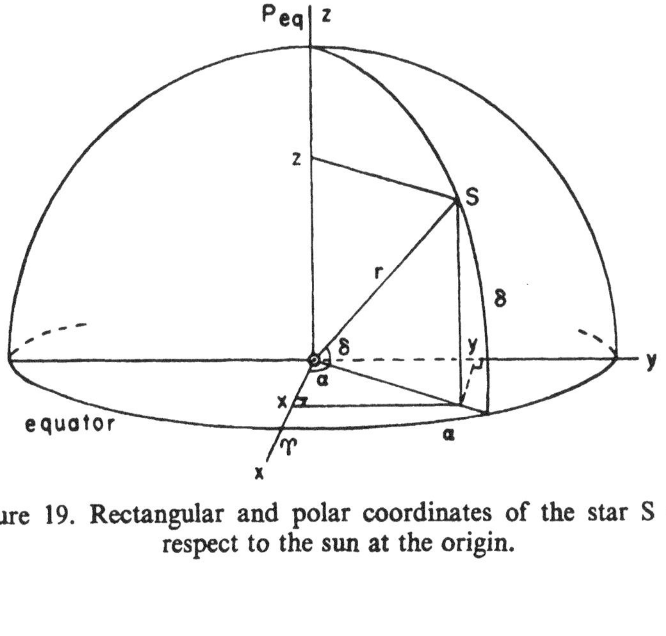

Figure 19. Rectangular and polar coordinates of the star S with respect to the sun at the origin.

We will call the velocity components u, v, w. Further the radial velocity $V_r = dr/dt$ and the proper motion components $\mu_\alpha = d\alpha/dt$ and $\mu_\delta = d\delta/dt$ are observed. We have then in the new notation:

$$\left.\begin{aligned} u &= V_r \cos\delta\cos\alpha - r\,\mu_\delta \sin\delta\cos\alpha - r\,\mu_\alpha \cos\delta\sin\alpha \\ v &= V_r \cos\delta\sin\alpha - r\,\mu_\delta \sin\delta\sin\alpha + r\,\mu_\alpha \cos\delta\cos\alpha \\ w &= V_r \sin\delta + r\,\mu_\delta \cos\delta \end{aligned}\right\} \quad (35)$$

Solving for V_r, μ_δ, $\mu_\alpha \cos\delta$ gives:

$$V_r = +u\cos\alpha\cos\delta + v\sin\alpha\cos\delta + w\sin\delta \quad (36)$$

$$\left.\begin{aligned} \mu_\delta &= -\frac{u}{r}\cos\alpha\sin\delta - \frac{v}{r}\sin\alpha\sin\delta + \frac{w}{r}\cos\delta \\ \mu_\alpha\cos\delta &= -\frac{u}{r}\sin\alpha + \frac{v}{r}\cos\alpha \end{aligned}\right\} \quad (37)$$

For a single star we do not know the peculiar space motion in advance. However, we can assume random motions of the stars. For all stars in one magnitude interval the peculiar radial velocities will cancel in the mean value. The same holds for the peculiar proper motions if we assume that all stars in the magnitude interval have the same distance to the sun. We will consider therefore further on the mean values of the stars concerned. The velocity components u, v, w are then the components of the reflected solar motion. From the radial velocity observations we see that we can derive the u, v, w by least squares, but from the proper motion observations we find only the u/r, v/r, w/r.

Let $V_\odot$ be the linear solar speed directed towards the apex with coordinates A, D. The speed $-V_\odot$ will have then the components u, v, w, as observed from the motions of the stars. From a similar figure we find a similar relation as formula (34).

$$\left.\begin{aligned} u &= -V_\odot \cos D\cos A \\ v &= -V_\odot \cos D\sin A \\ w &= -V_\odot \sin D \end{aligned}\right\} \quad (38)$$

Further we have:

$$u^2 + v^2 + w^2 = V_{\odot}^2 \tag{39}$$

$$\left.\begin{aligned} \tan A &= \frac{v}{u} = \frac{v/r}{u/r} \\ \tan D &= \frac{\sin D}{\cos D} = \frac{w}{\sqrt{(u^2+v^2)}} = \frac{w/r}{\sqrt{\{(u/r)^2+(v/r)^2\}}} \end{aligned}\right\} \tag{40}$$

From the radial velocity observations both the size $V_{\odot}$ and direction A, D of the apex can be found; from the proper motion observations only the direction or the position of the apex in the sky can be derived, as could be expected a priori. It can be seen from the last formulae that the units used are of no importance as long as they are consistent.

The results found from the nearby stars are $V_{\odot} = 20$ km/sec, $A = 18^h$, $D = +30°$. It should be noted here that we find the speed and direction of the relative solar motion with respect to the nearby stars. If we study more distant objects we find another relative speed and direction. Finally when we study the globular clusters we find the motion of our sun with respect to the center of the Milky Way system.

19. *Secular parallax*. We have seen that the sun moves through space towards the apex with a speed of about 20 kilometers per second with respect to the local standard of rest. The reflected space motion of a star is the same 20 kilometers per second directed towards the antapex. If the angular distance between antapex and star is called σ the radial velocity of the star will now be 20 cos σ kilometers per second as far as the reflection of the solar motion is concerned (Figure 20).

We have: one astronomical unit/year = 149.5 million km/ 31.56 million seconds = 4.74 km/sec. The sun moves 20/4.74 or 4.2 astronomical units through space per year relative to the nearby stars. In the tangential plane this reflected speed is 20 sin σ kilometers per second or 4.2 sin σ astronomical units per year. The proper motion component in the antapex direction depends in addition on the distance of the star. If the star had no peculiar motion, then this shift in the plane of the sky towards the antapex would determine the distance of the star. Unfortunately the peculiar motion is unknown for a single star but for a group of similar stars one can again assume that they cancel in the mean value assuming random motions of the stars. This gives us the secular parallax which is thus a statistical parallax.

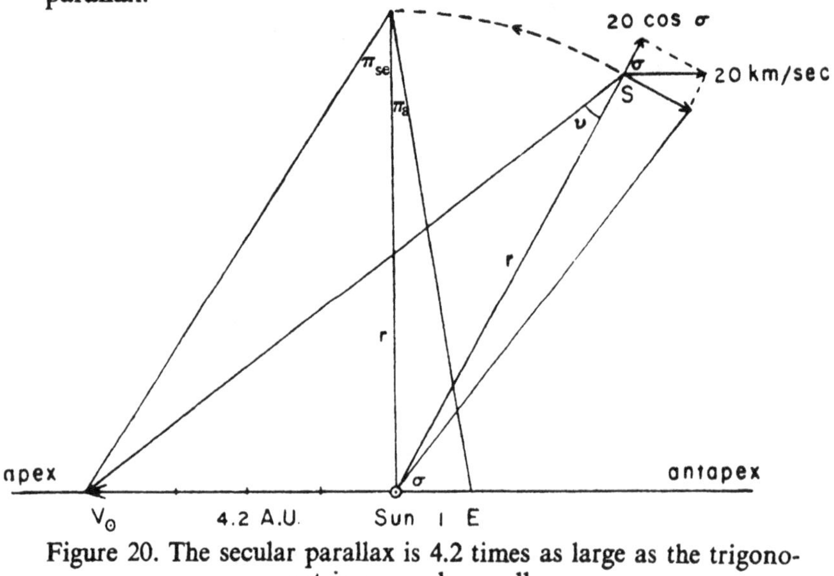

Figure 20. The secular parallax is 4.2 times as large as the trigonometric or yearly parallax.

The base line is 4.2 times as long per year as the distance sun-earth and moreover this base line increases throughout the years. If π_{se} means secular parallax, and π_a the absolute trigonometric or annual parallax, we have accordingly the following relation:

$$\pi_{se} = 4.2\,\pi_a \tag{41}$$

Call the component of the star's proper motion in the antapex direction υ and perpendicular to it τ (Figure 21). Then $\Sigma\tau = 0$. As before independent of the units used it follows:

$$\upsilon = \pi_{se} \sin\sigma = 4.2\,\pi_a \sin\sigma, \qquad \pi_{se} = \frac{\upsilon}{\sin\sigma} \tag{42}$$

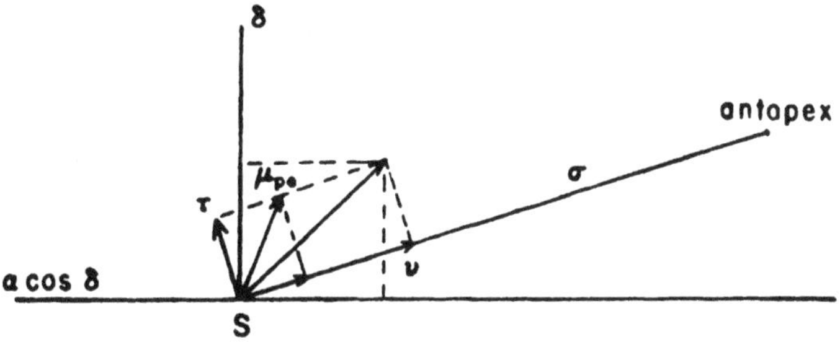

Figure 21. The observed proper motion μ of a star with the components υ, τ. The peculiar proper motion of the star may be μ_{pe}.

We can measure the υ while we can compute the σ. However, the weights are not the same for all stars, but are largest 90° off from the apex direction. There one sees the largest deflection of the solar motion in the proper motion of the star. If ϵ is the probable error then we measure $\upsilon \pm \epsilon$ thus weight $p = 1/\epsilon^2$.

$$\pi_{se} = \frac{\upsilon}{\sin\sigma} \pm \frac{\epsilon}{\sin\sigma}, \qquad p' = \frac{\sin^2\sigma}{\epsilon^2} = p\sin^2\sigma \tag{43}$$

If for all stars the measured proper motions are of equal accuracy, $\epsilon_1 = \epsilon_2 = \ldots$ and $p' = \sin^2\sigma$. Therefore :

$$\pi_{se} = \frac{\frac{\upsilon_1}{\sin\sigma_1}\sin^2\sigma_1 + \ldots}{\sin^2\sigma_1 + \ldots} = \frac{[\upsilon\sin\sigma]}{[\sin^2\sigma]} = \frac{\overline{\upsilon\sin\sigma}}{\overline{\sin^2\sigma}} \tag{44}$$

However, when the accuracy is not the same, which is more probable, we have $\epsilon_1 \neq \epsilon_2$ and find:

$$\pi_{se} = \frac{\dfrac{v_1}{\sin \sigma_1} p_1 \sin^2 \sigma_1 + \ldots}{p_1 \sin^2 \sigma_1} = \frac{v_1 p_1 \sin \sigma_1 + \ldots}{p_1 \sin^2 \sigma_1 + \ldots}$$

$$\pi_{se} = \frac{[vp \sin \sigma]}{[p \sin^2 \sigma]} = \frac{\overline{vp \sin \sigma}}{\overline{p \sin^2 \sigma}} \tag{45}$$

For deriving the yearly parallax we divide by 4.2 as seen above.

For a star twice as far away the reflection of the solar motion is twice as small in the proper motion. In the radial velocities it remains the same because this is an absolute measurement. For a certain spectral group of stars one can also say that the peculiar proper motions become twice as small if the distance becomes twice as large. Thus in general the proper motion is proportional to the parallax of the group.

REFERENCES

SPHERICAL ASTRONOMY

W. M. Smart: *Spherical astronomy*. Cambridge Un. Press, 1956.

A. Prey: *Einführung in die Sphärischen Astronomie*. Springer Verlag, Wien, 1949.

F. Becker: *Grundrisz der Sphärischen und Praktischen Astronomie*. Dümmlers Verlag, Berlin und Bonn, 1934.

LEAST SQUARES METHOD

D. Brunt: *The combination of observations*. Cambridge Un. Press, 1931.

E. und B. Strömgren: *Lehrbuch der Astronomie*. Springer Verlag, Berlin, p. 493, 1933.

TRIGONOMETRIC PARALLAX

W. M. Smart: *Foundations of astronomy*. Longmans, Green and Co., London, New York, Toronto, 1942.

PHOTOGRAPHIC STELLAR PARALLAX

F. Schlesinger: *Festschrift für Hugo v. Seeliger*, p. 422, 1924.

P. van de Kamp: *Pop. Astr.* **59**, 65, 129, 176, 243, 1951.

CORRECTIONS TO ABSOLUTE PARALLAX

A. N. Vyssotsky and E. T. R. Williams: *Publ. McCormick Obs.*, **10**, 33 and 36, 1948.

L. Binnendijk: *B.A.N.*, **10**, 9, 1943.

SOLAR MOTION AND APEX

M. Waldmeier: *Einführung in die Astrophysik*. Birkhäuser Verlag, Basel, 1948.

SECULAR PARALLAX

P. van de Kamp: *Ann. New York Acad. of Sci.*, **42**, 151, 1941.

E. T. R. Williams: *Ann. New York Acad. of Sci.*, **42**, 187, 1941.

CATALOGUE OF TRIGONOMETRIC PARALLAXES

L. F. Jenkins: *General catalogue of trigonometric stellar parallaxes*. Yale Un. Obs., New Haven, 1952.

CATALOGUES OF PROPER MOTIONS

Boss general catalogue., Vol. **1–5**, Carnegie Inst., 1937.

Cape zone catalogue., Cape Obs., 1936.

Radcliffe catalogue of proper motions in Selected Areas 1–115, Oxford, 1934.

Poulkovo catalogue of proper motions in Selected Areas, Poulkovo Publ., Series II, Vol. **55**, 1940.

First and second McCormick catalogue, Publ. McCormick Obs., **7**, 1937; **10**, 1948.

II

Visual Double Stars

VISUAL DOUBLE STARS FALL INTO TWO CATEGORIES: OPTICAL doubles in which the components are widely separated in space and appear together only because they lie nearly in the same line of sight, and physical double stars in which the components are each in orbital motion around the center of mass of the system. W. Herschel was the first to realize that there were many more physical double star systems than optical systems because he found many more of them than would agree with a random distribution. The older observations are now compiled in catalogues, for example:

β.G.C. = Burnham General Catalogue 1906; 14,000 systems
A.D.S. = Aitken New General Catalogue 1932; 17,000 systems

Together with the southern double stars the number of known double stars is about 23,000. Of these only about 150 have well determined orbits. The trouble is obviously that the periods are very long and the orbit has to be covered over at least half the period to give a good determination of the orbital elements.

We have to distinguish between the true spatial orbit and the apparent orbit which is projected on the plane of the sky. The orbital determination derives the elements of the true orbit from the observed apparent orbit.

20. *Observation.* The method of visual observation was standardized by F. J. W. Struve. One measures polar coordinates, namely the angular distance between the two stars in seconds of

arc and the position angle, namely the angle at the primary between the north direction and the direction towards the component, reckoned eastward. The direction east-west is found by daily motion of the star when the clockdrive is stopped.

There are two parallel wires in the micrometer, one fixed and one movable by means of a fine screw, whose position can be read by means of a divided head (Plate III). The instrument can be rotated around the optical axis. The position angle θ can be determined (Figure 22). To measure the angular distance ρ the primary star S_1 is placed in the center of the field bisecting the fixed wire and the movable wire is made to bisect the image of star S_2; next the micrometer box is shifted slightly and the star S_2 is placed in the center of the field bisecting the fixed wire thus displacing the movable wire. This last wire is made to pass over S_2 until it bisects the image of star S_1. In this way twice the

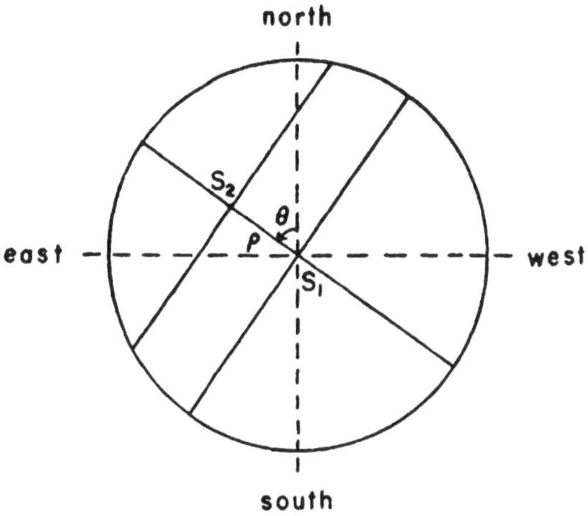

Figure 22. The polar coordinates of a visual double star are the angular distance ρ in seconds of arc and the position angle θ. The field of view of an astronomical telescope is inverted.

angular distance is obtained and the necessity of determining the coincidence of the wires is eliminated. The screw value is obtained from star transits or from known large angular distances. The resolving power is given by 4".56 divided by the aperture in inches.

Photographic observations will show two images on the plate only when the pair is a wide one. This method was standardized by E. Hertzsprung. If the components are rather bright the exposure time can be short and one can make a whole sequence of exposures on the photographic plate to gain accuracy. The shifting of the telescope in right ascension has to be done opposite the daily motion because the inertia of the telescope is in the direction of the daily motion (Plate IV). It is important that both stars have about the same magnitude. Often this is not the case. One uses then an objective grating consisting of a number of parallel bars in front of the objective. Two fainter diffraction or first order images S′ appear at the same distance on both sides from the central image S° on the photographic plate (Figure 23). Instead of measuring S_2° with respect to S_1° one measures S_2° with respect to the average values of the S_1'. The objective grating is chosen in such a way that the images

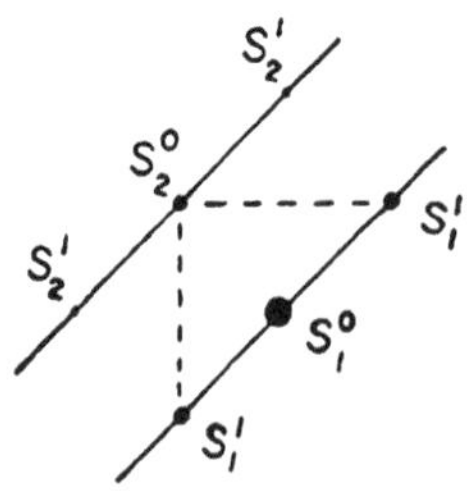

Figure 23. The grating images of a wide double star with components S_1 and S_2. Measured are the positions of S_2° and both S_1' in rectangular coordinates.

S_1' and S_2° have almost the same density on the plate. In addition a trail is made to define the east-west direction on the plate.

The measurement is carried out in much the same way as described already in paragraph 10. We again rotate our plate to a convenient equinox, $t_0 = 1950$ for example, and have then the equator direction for that year. We measure the rectangular coordinates of the images, rotate the plate by 180° and measure again.

21. *Orbital elements defined.* Seven elements define the orbit in the true plane. The first three are called dynamical elements and define the motion in the orbit; the last four are the geometrical elements and give the size and orientation of the

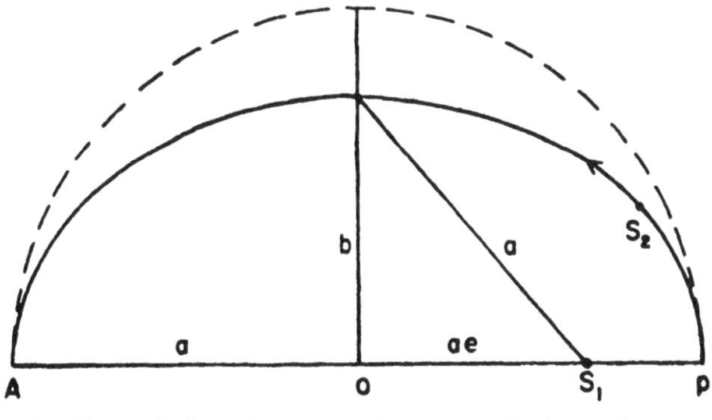

Figure 24. The relation between the axes of the ellipse and the eccentricity.

orbit. The elements are defined somewhat differently from the case of planets and comets:

P = period of revolution in mean solar years. The average yearly angular motion is $n = 2\pi/P$.

T = time of periastron passage.

e = eccentricity of the true ellipse (Figure 24).

a = semi-major axis of true ellipse.

Ω = position angle of the nodal point between 0° and 180°. It is also the position angle of the line of the nodes.

ω = the angular distance measured in the true orbit between the line of the nodes and periastron from 0° to 360° in the direction of the motion of the component (See in Figure 25).

i = the inclination of the orbital plane on the plane of the sky. For direct motion when the position angle increases with time the i is between 0° and 90°, for retrograde motion the i is counted between 90° and 180°. The inclination is defined as positive when at the nodal point the companion is moving away from us in relation to the principal star.

Observations of visual double stars do not give the sign of the inclination and therefore do not tell about the quadrant. There-

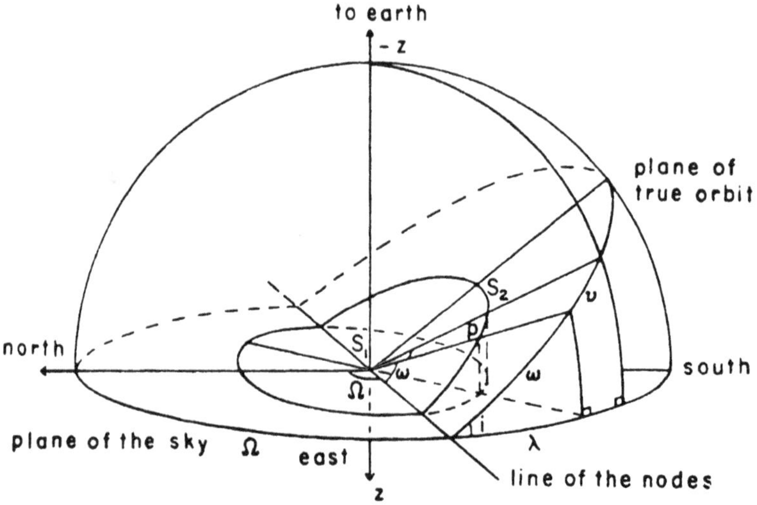

Figure 25. The true ellipse in space and the apparent ellipse in the plane of the sky.

fore one does not know which is the increasing and which the decreasing node. One takes that node which has a position angle smaller than 180° and calls it the nodal point. Radial velocity observations can determine the quadrant.

22. *Gravitation law and Kepler's laws.* Newton's general law of gravitation can be stated as follows: Every particle of matter in the universe attracts every other with a force which is proportional to the product of the masses and inversely proportional to the square of the distance between them.

$$F = G\,\frac{\mathfrak{M}_1 \mathfrak{M}_2}{r^2} \tag{46}$$

Here F = force, G = gravitational constant = 6.673×10^{-8} cm^3/gm. sec^2, $\mathfrak{M}$ = mass, r = distance between the masses.

Kepler's three laws follow mathematically from the gravitation law and hold therefore for a binary system in the true orbit.

(1) The absolute orbit of each component is an ellipse with the gravity center at one of its foci. The relative ellipse of the fainter component with respect to the brighter component is an ellipse with the brighter star at one of its foci. This ellipse is usually written in polar coordinates:

$$r = \frac{a(1 - e^2)}{1 + e \cos v} \tag{47}$$

Here: r = radius vector, e = eccentricity, v = true anomaly.

(2) *The law of areas.* The radius vector of each component sweeps equal areas in equal times. For the relative ellipse in the true orbit we have if h is a constant:

$$h = r^2\,\frac{dv}{dt} \tag{48}$$

The law holds under projection as well and is thus also good for the apparent ellipse, but we have to use another constant h'.

(3) The *harmonic law*. The sum of the masses of the components are equal to the cube of their mean distance divided by the square of their period. The units are solar mass, astronomical unit, and year. The elements a and P are thus dependent.

$$\mathfrak{M}_1 + \mathfrak{M}_2 = \frac{a^3}{P^2} \tag{49}$$

We have neglected here the mass of the earth. Strictly speaking the sidereal year or the true period of the earth's revolution should be taken as a unit. In practice one always takes the (calendar) year which should coincide with the tropical year. The ratio of sidereal year and tropical year is 1.00004, which is so nearly unity that we can neglect the distinction in double star astronomy.

23. *Apparent ellipse and true ellipse.* Because the apparent ellipse is the projection of the true ellipse we have several mathematical relations. The center of the true ellipse corresponds to the center of the apparent ellipse. The position of the bright star is the projection of one of the foci of the true ellipse (Plate V). The position of the bright star is therefore not at the focus of the apparent ellipse. The diameter of the apparent ellipse through the bright star is the projection of the major axis of the true ellipse; the extremes are periastron and apastron. A line parallel to the line of the nodes remains parallel and unshortened in the apparent ellipse. A line perpendicular to the line of the nodes remains perpendicular, but its length is shortened in the ratio cos i in the apparent ellipse.

One can draw the apparent ellipse through the observed normal points in several ways. One way is to do it analytically. An ellipse is determined by five elements; thus five points are enough. In this method one does not make use of the observing time. This is by far the most accurate observing result, and therefore it is better to use it.

One can also construct an ellipse through the normal points with help of an ellipsograph and check the law of areas with a planimeter. One changes the ellipse till the law is satisfied. Here the time observations are used.

Still another method makes use of the fact that the time is most accurately known. We plot the distance ρ in seconds of arc against the time. In the same graph we plot the position angle in degrees against the time (Figure 26). Smooth curves are drawn. The deviation of the normal points is in the ordinate direction only and gives a good idea of the accuracy of both observed quantities. We can now also plot the yearly change in the position angle against the time. This yearly change is found in degrees. By dividing by 57°29 we find it in radians. Now we can compute for any time the constant of areas, which is twice the amount of the yearly sector. We have thus:

$$h' = \rho^2 \frac{d\theta}{dt} \tag{50}$$

Here θ is in radians. We have to change our smooth curves until we find for this product the same constant h' at any time.

It is very important to draw a careful apparent ellipse, in which the law of areas is satisfied because we use this ellipse in the derivation of the orbital elements. It is now very easy from the two plots to draw the apparent ellipse.

24. *Method of M. Kowalsky and S. Glasenapp.* This old method is elegant in a mathematical way but no longer used in practice except for the determination of the P and T. The derivation of the formulae is given in Aitken's textbook *The Binary Stars*. We will give therefore only the underlying ideas and the final

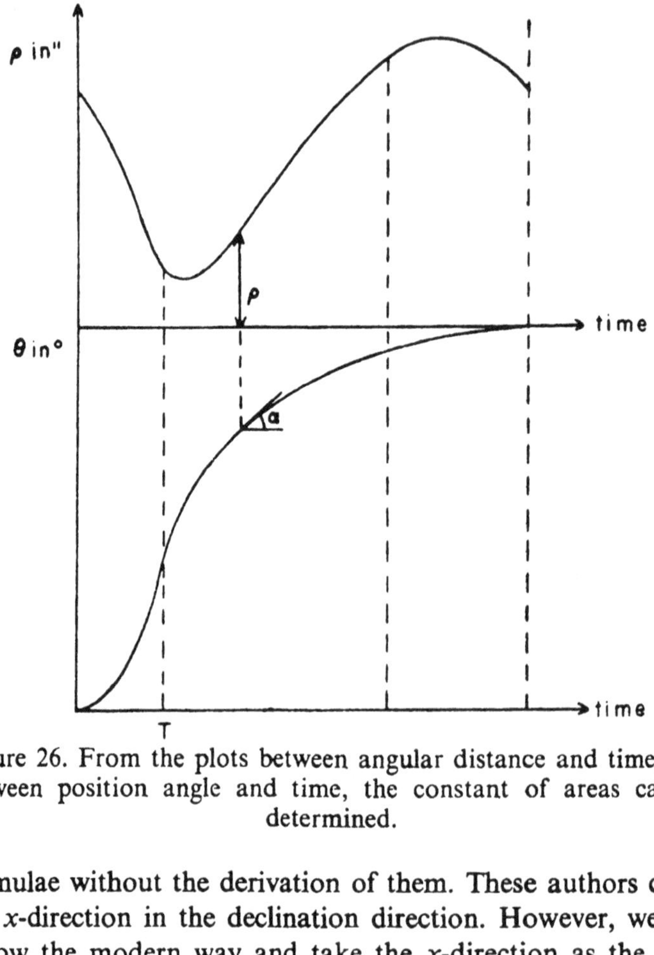

Figure 26. From the plots between angular distance and time, and between position angle and time, the constant of areas can be determined.

formulae without the derivation of them. These authors chose the x-direction in the declination direction. However, we will follow the modern way and take the x-direction as the right ascension direction and the y-direction as the declination direction. Our figures are drawn as seen by the naked eye in the southern sky at northern latitudes. The equation of the apparent ellipse is:

$$Ax^2 + 2Hxy + By^2 + 2Gx + 2Fy + 1 = 0 \qquad (51)$$

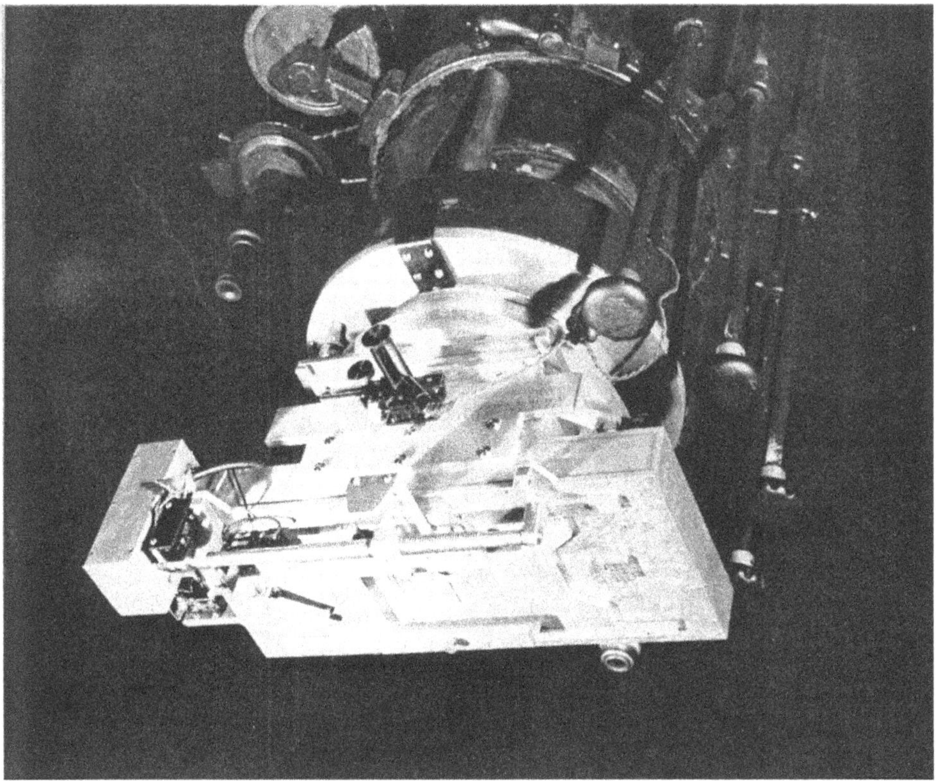

Plate IV. Double star camera attached to the Dearborn 18½-inch refractor. The camera allows for multiple exposures (35 in a row) with automatic timing and plate transport. The electronic timing unit (not shown in the picture) provides the necessary signals for operation of the plate holder, the camera shutter, and the timing of the exposures in the proper sequence of operation.
(Courtesy Dr. K. Aa. Strand.)

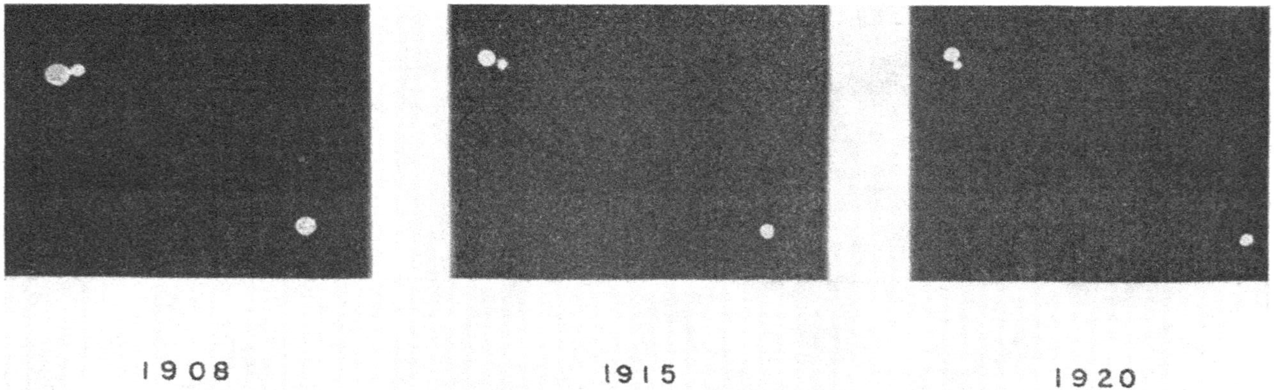

Plate V. Photographs of the visual binary system Krueger 60, in the upper left corner, show changes in position of the components. (Photographed by Barnard at Yerkes Observatory.)

Here we have five constants which will be determined by five points on the ellipse (Figure 27). We take first the two pairs of intersection with both axes and choose then a convenient fifth point. For $y = 0$ we find: $Ax^2 + 2Gx + 1 = 0$ and thus:

$$x_1 + x_2 = -\frac{2G}{A}, \qquad x_1x_2 = \frac{1}{A} \tag{52}$$

From this we determine the constants:

$$A = \frac{1}{x_1x_2}, \qquad G = -\frac{x_1 + x_2}{2x_1x_2}$$

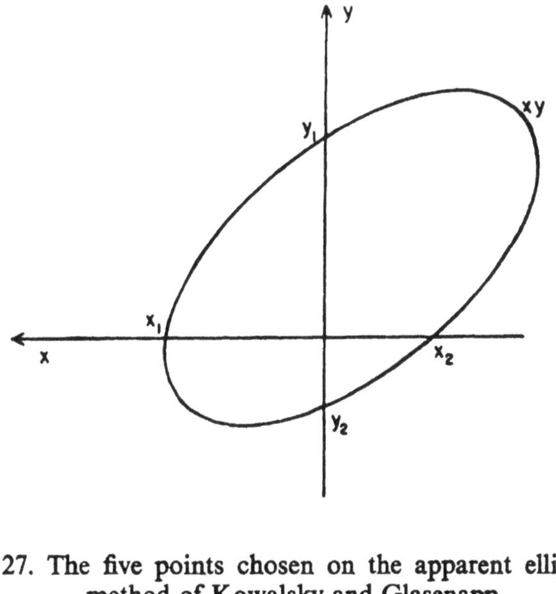

Figure 27. The five points chosen on the apparent ellipse in the method of Kowalsky and Glasenapp.

For $x = 0$ we find similar expressions for B and F. To determine H we choose a point where the product xy is large.

Seen in space formula (51) represents also a cylinder with

the axis in the line of sight (Figure 28). Take now a coordinate system with the same origin, the ξ-axis according to the line of nodes, the η-axis 90° different in the true plane, and the ζ-axis perpendicular to this plane. The transformation formulae are rather simple and include sine and cosine of the arguments i and Ω. We substitute now $\zeta = 0$ and get then the intersection of the cylinder with the true plane, namely the true ellipse.

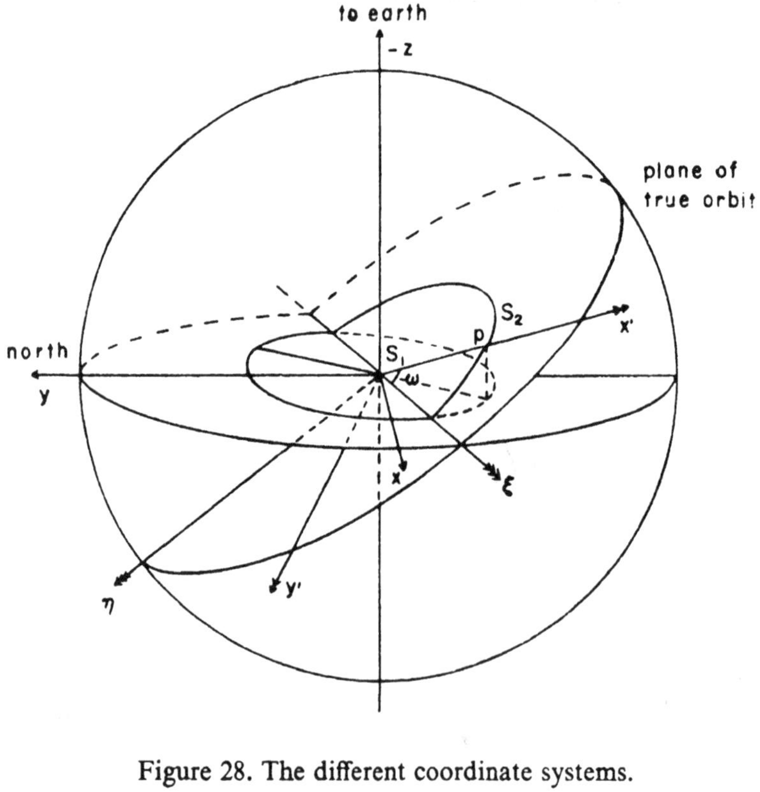

Figure 28. The different coordinate systems.

We can find another expression for the true ellipse. Let the x'-axis coincide with the major axis of the true ellipse and the position of the bright star at one of the focal points be taken as origin. The equation of the true ellipse is:

$$\frac{(x' + ae)^2}{a^2} + \frac{(y')^2}{b^2} = 1 \tag{53}$$

We turn now this axis back through the angle ω to make it coincide with the line of the nodes. After substituting thus the well known rotation formulae we will get the expression for the true ellipse on the same axes and the same origin.

Both expressions are identical therefore and the coefficients have to be proportional. This gives us six equations, from which one can find five orbital elements and the proportionality factor f. The final six equations can be written as:

$$f = \frac{1}{e^2 - 1} = -\frac{a}{p}, \qquad p = a(1 - e^2) \tag{54}$$

$$\frac{\tan^2 i}{p^2} \sin 2\Omega = 2(H - FG) \tag{55}$$

$$\frac{\tan^2 i}{p^2} \cos 2\Omega = G^2 - F^2 + B - A \tag{56}$$

$$\frac{2}{p^2} + \frac{\tan^2 i}{p^2} = F^2 + G^2 - (A + B) \tag{57}$$

$$\frac{e}{p} \sin \omega = +(F \sin \Omega - G \cos \Omega) \cos i \tag{58}$$

$$\frac{e}{p} \cos \omega = -(G \sin \Omega + F \cos \Omega) \tag{59}$$

Division of formula (55) by (56) gives:

$$\tan 2\Omega = \frac{2(H - FG)}{G^2 - F^2 + B - A} \tag{60}$$

Thus we find $\Omega \leqslant 180°$ and $(\tan^2 i)/p^2$ is also known from formulae (55) or (56). Substituting in formula (57) gives $2/p^2$

and thus p. Substitute now this p, then follows $\tan^2 i$ to find $\pm i$. We cannot find the sign of i. Division of formula (58) by (59) gives:

$$\tan \omega = \frac{G \cos \Omega - F \sin \Omega}{G \sin \Omega + F \cos \Omega} \cos i \tag{61}$$

This gives ω. We can now find e from formula (59) and a from formula (54). Thus in this method we measured five points on the apparent ellipse and found successively Ω, $\pm i$, ω, e, a.

In a graphical way we can determine P and T as follows. The areal velocity according to Kepler's second law has been determined. For the whole apparent ellipse it yields: 2 area $= Ph'$. If we determine the area with a planimeter, the P follows. One can also compute the area of the apparent ellipse by measuring the axes.

Periastron is one of the intersections between the apparent ellipse and the line joining the center and the bright star, and is closest to this star. The problem is to find the time when the comparison star is at that point of its orbit. Take the two normal points on each side of the periastron point and let the times be t_1 and t_2. For the areas s_1 and s_2 according to Kepler's second law we have:

$$\frac{s_1}{s_2} = \frac{T - t_1}{t_2 - T}, \qquad T = \frac{s_1 t_2 + s_2 t_1}{s_1 + s_2} \tag{62}$$

The areas can be determined with help of a planimeter. We can also take more normal points and take the average value as the best T.

Analytically we determine P and T as follows with help of the formulae of the two body problem. (For proof see paragraph 40 of this chapter.) Let be: v = true anomaly, E = eccentric anomaly, M_a = mean anomaly.

$$\tan (v + \omega) = \tan (\theta - \Omega) \sec i \tag{63}$$

$$\tan\frac{v}{2} = \sqrt{\frac{1+e}{1-e}}\tan\frac{E}{2} \tag{64}$$

$$M_a = E - e\sin E = \frac{2\pi}{P}(t - T) = n(t - T) \tag{65}$$

The last equation is known as Kepler's equation. From the first equation we find v for the normal point θ, from the second one we find E, and from the third one M_a. A second normal point gives another M_a. From both we can find the values P and T. In practice we use all normal points in a least squares solution.

25. *Method of H. J. Zwiers and H. N. Russell.* Graphically this method proceeds as follows. Let S_1 be the primary star and O the center of the apparent relative ellipse. Then p is the periastron (Figure 29). Since ratios are unaltered by projection we

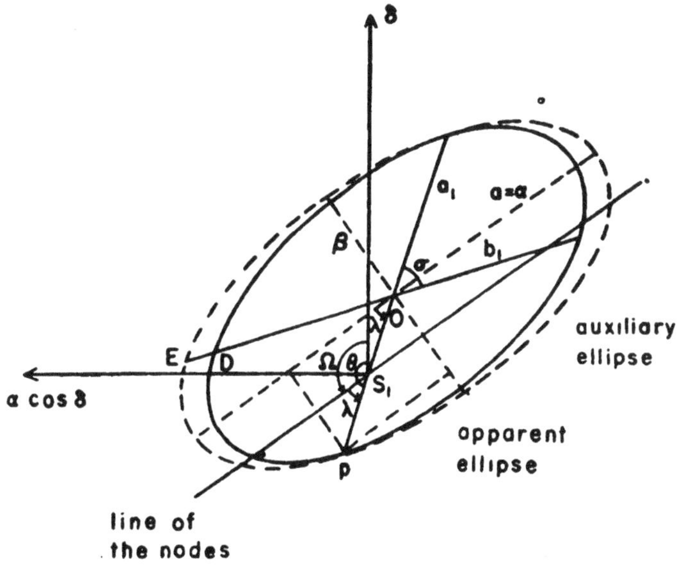

Figure 29. Apparent ellipse and auxiliary ellipse in the method of Zwiers and Russell.

find the eccentricity as a ratio of lines. $e = OS_1/Op$. By drawing any chord parallel to a_1 and taking the bisecting point, we can easily draw the conjugate diameter b_1. The ratio between semi-major and semi-minor axes is now:

$$k = \frac{a}{b} = \frac{1}{\sqrt{(1 - e^2)}} \leqslant 1 \qquad (66)$$

Now increase by this factor k in a direction parallel to b_1 all coordinates of the points of the relative ellipse. We get an auxiliary ellipse, which is the projection of the auxiliary circle around the true orbit. Now let α, β be the axes of this auxiliary ellipse. For the major axis we have $a = \alpha$, in seconds of arc, because the semi-major axis of the true ellipse equals the radius of the auxiliary circle and the major axis of the auxiliary ellipse is the only chord not shortened by projection. The perpendicular radius in the auxiliary circle is shortened to β by projection so that: $\cos i = \beta/\alpha$. This gives $\pm i$.

The diameter α was not altered by projection; hence the line of the nodes is parallel to this line and goes through star S_1. The position angle of the node Ω can now be read as the position angle of this constructed line. Also the position angle θ of periastron can be measured since $\lambda = |\theta - \Omega|$ is known. This λ is the projection of the angle ω, and the relation between the two is (See Figure 25):

$$\tan \lambda = \tan \omega \cos i \qquad (67)$$

At present we measure ω and thus also λ in the direction of the orbital motion. Therefore the absolute signs were introduced which were not given in the original method. We measured here thus OS_1, a_1, α, β, λ and found e, a, $\pm i$, Ω, ω.

Analytically we do not draw the auxiliary ellipse at all but make use of its properties. Let $OE = b_2 = k\, OD = kb_1$. The axes of the conjugate diameters of the apparent ellipse are thus

a_1, b_1 and the axes of the auxiliary ellipse α, β. According to the theorems of Apollonius we have if σ is the angle between a_1 and b_1 (For proof see paragraph 40 of this chapter):

$$\left.\begin{aligned} \alpha^2 + \beta^2 &= a_1^2 + b_2^2 &&= a^2_1 + k^2 b_1^2 \\ \alpha\,\beta &= a_1 b_2 \sin\sigma &&= k a_1 b_1 \sin\sigma \end{aligned}\right\} \tag{68}$$

From this we find now:

$$\left.\begin{aligned} (\alpha + \beta)^2 &= a_1^2 + 2k a_1 b_1 \sin\sigma + k^2 b_1^2 \equiv g^2 \\ (\alpha - \beta)^2 &= a_1^2 - 2k a_1 b_1 \sin\sigma + k^2 b_1^2 \equiv h^2 \end{aligned}\right\} \tag{69}$$

Because a_1, b_1, σ can be measured in the apparent ellipse and we know k, we can find the g and h, so that α and β are known.

$$a = \alpha = \tfrac{1}{2}(g + h), \qquad \beta = \tfrac{1}{2}(g - h) \tag{70}$$

In the same way as above we derive now cos i from which $\pm i$ results.

$$\cos i = \frac{\beta}{\alpha} = \frac{g - h}{g + h} \tag{71}$$

The coordinates of periastron with respect to the α and β are $a_1 \cos\lambda$ and $a_1 \sin\lambda$, where λ is again the angle between a_1 and α. This periastron point is a point on the auxiliary ellipse, thus has to satisfy the equation of this ellipse.

$$\frac{a_1^2 \cos^2\lambda}{\alpha^2} + \frac{a_1^2 \sin^2\lambda}{\beta^2} = 1 = \sin^2\lambda + \cos^2\lambda$$

$$\left(\frac{a_1^2}{\beta^2} - 1\right)\sin^2\lambda = \left(1 - \frac{a_1^2}{\alpha^2}\right)\cos^2\lambda$$

$$\frac{a_1^2 - \beta^2}{\beta^2}\sin^2\lambda = \frac{\alpha^2 - a_1^2}{\alpha^2}\cos^2\lambda$$

$$\tan^2\lambda = \frac{(\alpha^2 - a_1^2)}{\alpha^2}\,\frac{\beta^2}{(a_1^2 - \beta^2)} = \frac{\alpha^2 - a_1^2}{a_1^2 - \beta^2}\cos^2 i$$

The relation between λ and ω was: $\tan \lambda = \tan \omega \cos i$, so that:

$$\tan^2\omega = \frac{a^2 - a_1^2}{a_1^2 - \beta^2} \tag{72}$$

This gives $\pm\omega$. If we measure the position angle of the periastron, we have $\lambda = |\,\theta - \Omega\,|$. Because of the definition of Ω only one of the values of λ and ω suffices, and we find ω.

In the apparent ellipse we have to draw thus the conjugate diameters and measure a_1, b_1, OS_1, σ, θ_p. We find e, a, $\pm i$, Ω, ω. The P and T are found according to the previous method. To get still more accurate values we can apply differential corrections according to the method of least squares as given by G. C. Comstock.

26. *Method of T. N. Thiele and R. T. A. Innes.* This method starts with rectangular coordinates as measured on the photographic plate. The e is found according to the method of Zwiers and Russell, the P and T according to the method of Kowalsky and Glasenapp. The four remaining elements a, i, ω, Ω have to be found. Thiele and Innes choose the so-called "natural elements". Thiele introduced polar coordinates, but we will follow the rectangular coordinates of R. T. A. Innes and W. H. van de Bos. These authors took the declination direction as the x-axis. Again we will follow the modern way, namely the x-axis in the right ascension direction and the y-axis in the declination direction. The constants are those used by the authors.

The natural elements are A, B, F, G, which we will find to be the coordinates of periastron p (B, A) and the intersection of the conjugate diameter with the auxiliary ellipse q (G, F). The center of the apparent ellipse is taken here as origin. See Figure 30. If we draw the auxiliary ellipse, we can read off provisional values of *A*, *B*, *F*, *G* as the coordinates of two points on the ellipse. The unit is second of arc.

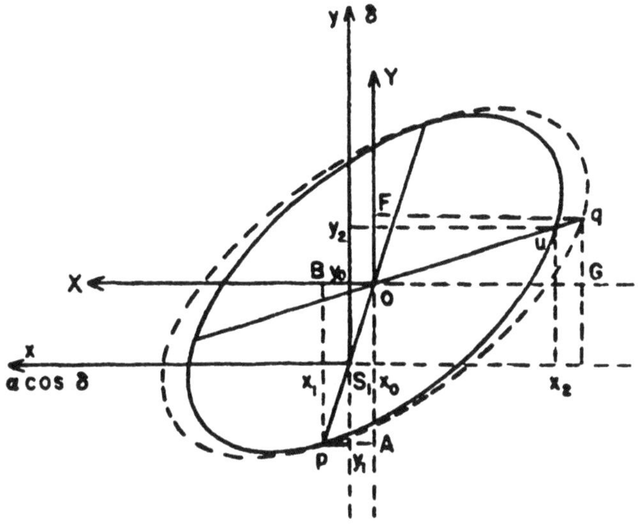

Figure 30. The constants in the method of Thiele and Innes are the coordinates of p and q with respect to the center of the apparent ellipse.

Let the coordinates as measured on the plate with respect to S_1 be $O(x_0, y_0)$, $p(x_1, y_1)$, $u(x_2, y_2)$. Then:

$$e = \frac{OS_1}{Op} = \sin\varphi = \frac{x_0}{x_0 - x_1} = \frac{y_0}{y_0 - y_1} \tag{73}$$

Thus e can be found. Later on we will correct this e to find a better eccentricity. From the introduction of the angle φ we find:

$$\cos^2\varphi = 1 - \sin^2\varphi = 1 - e^2, \quad \cos\varphi = \sqrt{(1 - e^2)}$$

$$k = \sec\varphi = \frac{1}{\sqrt{(1 - e^2)}} \tag{74}$$

The pair of coordinates can now be found from the measurements.

$$\left.\begin{aligned} A &= y_1 - y_0 = \delta_1 - \delta_0 \\ B &= x_1 - x_0 = a_1 \cos \delta_1 - a_0 \cos \delta_0 \end{aligned}\right\} \quad (75)$$

$$\left.\begin{aligned} F &= (y_2 - y_0)k = (y_2 - y_0) \sec \varphi \\ G &= (x_2 - x_0)k = (x_2 - x_0) \sec \varphi \end{aligned}\right\} \quad (76)$$

We use here only the coordinates of two points, but have to know also the coordinates of the center, which is not accurately determined. Those results are thus provisional.

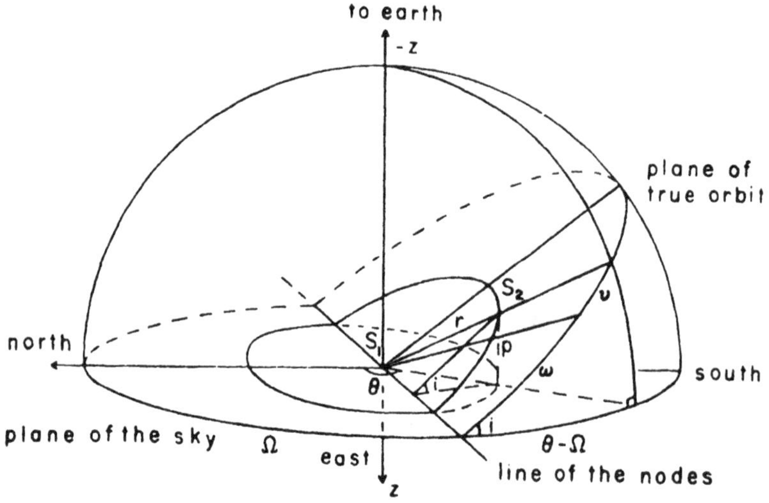

Figure 31. The space relation between true and apparent ellipse.

Analytically we proceed as follows (Figure 31). The ρ and θ are related to the usually used elements:

$$\left.\begin{aligned} \delta &= \rho \cos \theta \\ \alpha \cos \delta &= \rho \sin \theta \end{aligned}\right\} \quad (77)$$

$$\left.\begin{aligned} \rho \cos (\theta - \Omega) &= r \cos (v + \omega) \\ \rho \sin (\theta - \Omega) &= r \sin (v + \omega) \cos i \end{aligned}\right\} \quad (78)$$

Writing this out gives us:

$$\begin{aligned} \rho\cos\theta\cos\Omega + \rho\sin\theta\sin\Omega &= r\cos v\cos\omega - r\sin v\sin\omega \\ \rho\sin\theta\cos\Omega - \rho\cos\theta\sin\Omega &= r\sin v\cos\omega\cos i \\ &\quad + r\cos v\sin\omega\cos i \end{aligned}$$

First we multiply respectively by cos Ω and $(-\sin\ \Omega)$, and add.

$$\left.\begin{aligned} \delta \quad &= r\cos v(+\cos\omega\cos\Omega - \sin\omega\sin\Omega\cos i) \\ &\quad + r\sin v(-\sin\omega\cos\Omega - \cos\omega\sin\Omega\cos i) \\ a\cos\delta &= r\cos v(+\cos\omega\sin\Omega + \sin\omega\cos\Omega\cos i) \\ &\quad + r\sin v(-\sin\omega\sin\Omega + \cos\omega\cos\Omega\cos i) \end{aligned}\right\} \quad (79)$$

The last line we found by multiplying respectively by sin Ω and cos Ω and adding.

Let us write now, knowing only the dynamical elements P, T, e:

$$\left.\begin{aligned} x' &= \frac{r}{a}\cos v = \cos E - e \\ y' &= \frac{r}{a}\sin v = \sqrt{(1-e^2)}\sin E \end{aligned}\right\} \quad (80)$$

These are nothing other than the coordinates of the unit orbital ellipse in the true plane. We thus take $a = 1$ (Figure 32). For the given e they can be computed as soon as E is known.

$$M_a = E - e\sin E = \frac{2\pi}{P}(t - T)$$

M_a follows from the P, T, t and thus also E. Thus we can make a table in the following form, as was published in *Union Circ.* No. 71.

$$x', y' = f(e, M_a)$$

For the given e we compute the M_a for the normal point and find x', y'. So we can find the entire unit orbital ellipse.

Let us define further:

$$\left.\begin{aligned} A &= a(+\cos\omega\cos\Omega - \sin\omega\sin\Omega\cos i) \\ B &= a(+\cos\omega\sin\Omega + \sin\omega\cos\Omega\cos i) \\ F &= a(-\sin\omega\cos\Omega - \cos\omega\sin\Omega\cos i) \\ G &= a(-\sin\omega\sin\Omega + \cos\omega\cos\Omega\cos i) \end{aligned}\right\} \quad (81)$$

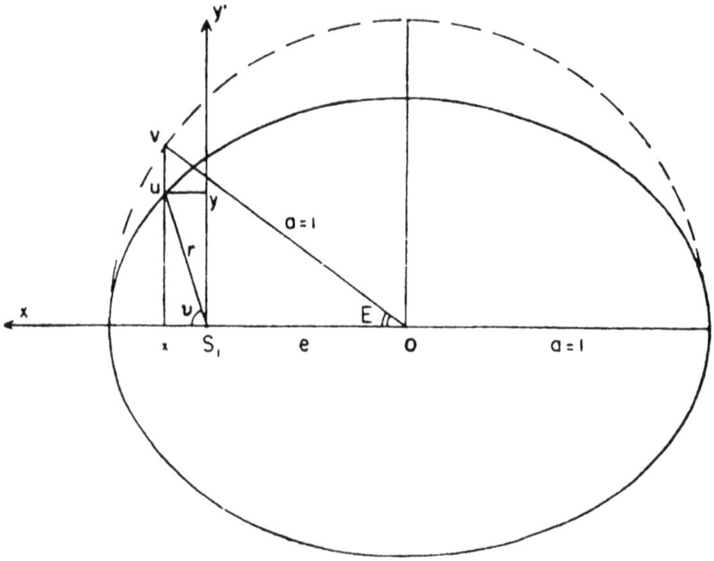

Figure 32. The unit orbital ellipse for a given eccentricity.

The A, B, C are the projections of the center-periastron distance in the true ellipse onto the three coordinate axes. In the same way F, G, H are the projections of the distance between center and the intersection of the minor axis and auxiliary circle in the true ellipse. Later we will introduce C and H, which are distances in the line of sight. We will prove this statement for one of the constants, namely B along the right ascension direction. The periastron has point B as projection on the right ascension axis. Now project the distance center-periastron on the line through the center, which is parallel to the line of the nodes. This gives ($a \cos \omega$), which gives ($a \cos \omega \sin \Omega$) after projection on the right ascension axis. We also see in Figure 33 that ($a \sin \omega$) is projected through angle i and thus gives ($a \sin \omega \cos i$) in the plane of the sky. It gives ($a \sin \omega \cos i \cos \Omega$) as projection on the right ascension axis. The figure shows that the signs are correct.

The A/a, B/a, C/a and F/a, G/a, H/a are the direction cosines of the semi-major and semi-minor axes of the true ellipse with respect to the three coordinate axes.

The previously mentioned formula (79) for the coordinates of a measured normal point can now be written as:

$$\left.\begin{aligned} y &= \delta &&= Ax' + Fy' \\ x &= \alpha \cos \delta &&= Bx' + Gy' \end{aligned}\right\} \qquad (82)$$

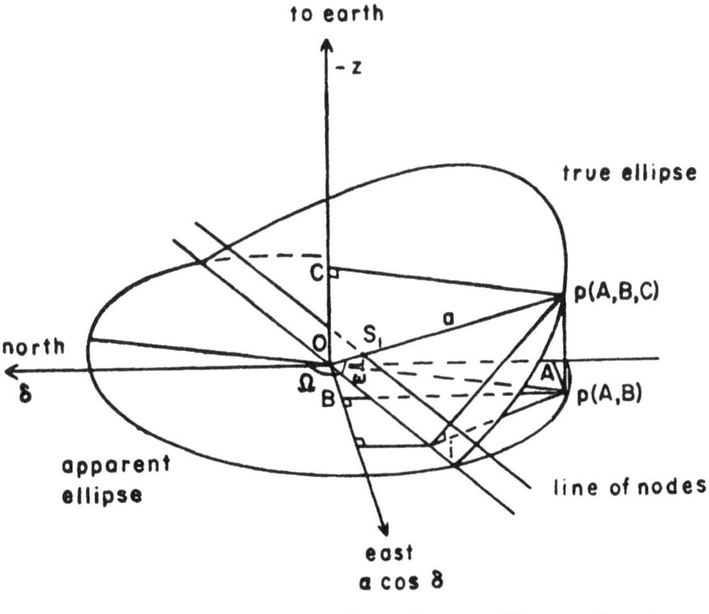

Figure 33. The constants A, B, C are the coordinates of periastron.

The observed positions in right ascension and declination are thus linear functions of the coordinates x' and y' of the unit ellipse in the true plane. The coefficients A, B, F, G determine the size, orientation and projection. The problem is how to determine those constants.

For the declination measurement of one normal point we get:

$$F = -\frac{x'}{y'} A + \frac{\delta}{y'}$$

This is a straight line. For the declination measurement of a second normal we get another straight line with a similar formula (Figure 34). They do not coincide because the δ, x', y' are different. The intersection gives the A, F which suffice as the roots of the two equations. The right ascension values for the

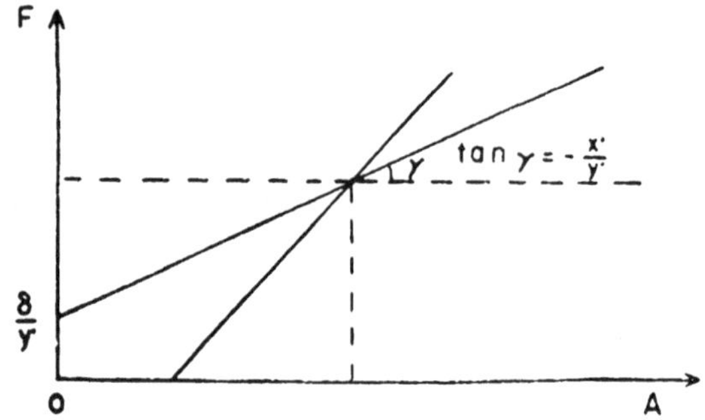

Figure 34. The constants A and F can be determined from the declination values of two normal points; the constants B and G from the right ascension values.

two same normal points give the values of the constants B, G. We can also draw the straight lines for the other normal points. They will all pass close to a certain point in the neighborhood of our first intersection. The coordinates of this point define the best A, F and similarly B, G. In practice we do this with the least squares solutions of each of our two equations which give also the probable errors of the constants A, B, F, G. Both the co-ordinates of all points are used therefore. If we divide by a we get the coordinates of the unit ellipse in the plane of the sky.

To convert the elements A, B, F, G, into the usual ones we write:

$$\begin{aligned} A + G &= a\cos(\omega + \Omega) + a\cos(\omega + \Omega)\cos i \\ &= a\cos(\omega + \Omega)(1 + \cos i) \\ &= 2a\cos(\omega + \Omega)\cos^2 \tfrac{1}{2}i \end{aligned}$$

We can find the different combinations:

$$\begin{aligned} A + G &= 2a\cos(\omega + \Omega)\cos^2 \tfrac{1}{2}i \\ A - G &= 2a\cos(\omega - \Omega)\sin^2 \tfrac{1}{2}i \\ B - F &= 2a\sin(\omega + \Omega)\cos^2 \tfrac{1}{2}i \\ -B - F &= 2a\sin(\omega - \Omega)\sin^2 \tfrac{1}{2}i \end{aligned}$$

From this we get the following formulae:

$$\left.\begin{aligned} \tan(\omega + \Omega) &= \frac{B - F}{A + G} \\ \tan(\omega - \Omega) &= \frac{B + F}{G - A} \end{aligned}\right\} \qquad (83)$$

This gives $(\omega + \Omega)$, $(\omega - \Omega)$, thus ω and Ω.
Further we can find $\pm i$ from:

$$\tan^2 \tfrac{1}{2}i = \frac{(A-G)\cos(\omega + \Omega)}{(A+G)\cos(\omega - \Omega)} = \frac{(B+F)\sin(\omega + \Omega)}{(F-B)\sin(\omega - \Omega)} \qquad (84)$$

We can find a in seconds of arc from the expressions for one of the four constants A, B, F, G in formula (81). We can also compute four values and take the average result as the best value. However, there is a more elegant method in which we use all four constants. The axes in the auxiliary ellipse are $a = \alpha$ and

$a\cos i = \beta$. According to the theorems of Apollonius we have:

$$a^2 + a^2\cos^2 i = \mathrm{p}O^2 + \mathrm{q}O^2 = (A^2 + B^2) + (F^2 + G^2) = 2u$$

$$a\,.\,a\cos i = \begin{vmatrix} A & B \\ F & G \end{vmatrix} = AG - BF = v$$

Thus we have now:

$$a^2(1 + \cos^2 i) = a^2\left(1 + \frac{v^2}{a^4}\right) = a^2 + \frac{v^2}{a^2} = 2u$$

$$a^4 - 2ua^2 + v^2 = 0 \tag{85}$$

The roots of this equation are:

$$a^2 = \frac{2u \pm \sqrt{(4u^2 - 4v^2)}}{2} = u + \sqrt{(u + v)(u - v)} \tag{86}$$

The negative sign does not suffice because then a^2 would be smaller than u. The first equation shows that the value of a^2 lies between u and $2u$.

We can check the P and T in the following way. We had:

$$\left.\begin{aligned} Ax' + Fy' &= y \\ Bx' + Gy' &= x \end{aligned}\right\}$$

The roots of the two equations are found from:

$$x' = \frac{\begin{vmatrix} y & F \\ x & G \end{vmatrix}}{\begin{vmatrix} A & F \\ B & G \end{vmatrix}} = \frac{Gy - Fx}{AG - BF}, \quad y' = \frac{\begin{vmatrix} A & y \\ B & x \end{vmatrix}}{\begin{vmatrix} A & F \\ B & G \end{vmatrix}} = \frac{Ax - By}{AG - BF} \tag{87}$$

By substituting the observed values x, y we find the x', y' for our normal point. In the table of *Union Circ.* we find then a value M_a for each normal point, which has to be linear with the time. $M_a = (t - T)\,2\pi/P$. For $M_a = 0$ we find T (Figure 35). The tangent of the slope gives $2\pi/P$ and thus P. A least squares solution can be made, giving also the probable errors in P and T.

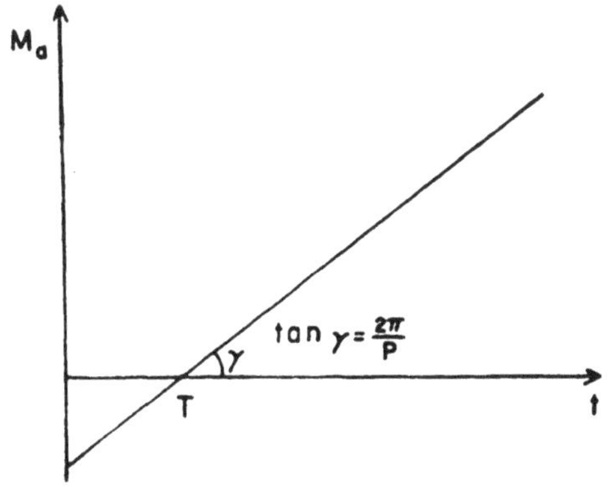

Figure 35. The relation between mean anomaly and time gives the time of periastron passage T and the period P.

27. *Differential corrections.* The differential corrections in this case are very simple and we will give them therefore as an example. We will again call $n = 2\pi/P$.

$$\left.\begin{aligned} \Delta y &= x' \, \Delta A + y' \, \Delta F + P_y \, \Delta e + Q_y n \, \Delta T + R_y \, \Delta n \\ \Delta x &= x' \, \Delta B + y' \, \Delta G + P_x \, \Delta e + Q_x n \, \Delta T + R_x \, \Delta n \end{aligned}\right\} \quad (88)$$

In each equation are 5 of the 7 elements. We have:

$$\left.\begin{aligned} P_y &= +A \frac{dx'}{de} + F\frac{dy'}{de}, & P_x &= +B\frac{dx'}{de} + G\frac{dy'}{de} \\ Q_y &= -A \frac{dx'}{dM_a} - F\frac{dy'}{dM_a}, & Q_x &= -B\frac{dx'}{dM_a} - G\frac{dy'}{dM_a} \\ R_y &= -(t - T)Q_y, & R_x &= -(t - T)Q_x \end{aligned}\right\} \quad (89)$$

The e, P, T were found from the drawing. By least squares we now determine the corrections to those elements and also those to the four constants A, B, F, G. This gives the best possible elements derived from the observations.

28. *Application to the line of sight.* The same notation can be used for the third coordinate, namely in the line of sight. This

gives the radial velocity after differentiating with respect to time.

$$z = r \sin (v + \omega) \sin i \tag{90}$$
$$= r \sin v \cos \omega \sin i + r \cos v \sin \omega \sin i$$

Let us call:

$$\left.\begin{array}{l} C = a \sin i \sin \omega \\ H = a \sin i \cos \omega \end{array}\right\} \tag{91}$$

Then we find:

$$z = Cx' + Hy' \tag{92}$$

For the direction cosines of the major axis and those of the minor axis the following three relations hold:

$$\left.\begin{array}{l} A^2 + B^2 + C^2 = a^2 \\ F^2 + G^2 + H^2 = a^2 \\ AF + BG + CH = 0 \end{array}\right\} \tag{93}$$

The last equation states that both axes are perpendicular in space. From the given A, B, F, G we can find now also the C and H with help of those relations. If we adopt the sign of C we have that of H.

$$M_a = \frac{2\pi}{P}(t - T), \qquad \frac{dM_a}{dt} = \frac{2\pi}{P} = n \tag{94}$$

We find now:

$$V_r = \frac{dz}{dt} = C\frac{dx'}{dt} + H\frac{dy'}{dt} = C\frac{dM_a}{dt}\frac{dx'}{dM_a} + H\frac{dM_a}{dt}\frac{dy'}{dM_a}$$
$$V_r = n\left(C\frac{dx'}{dM_a} + H\frac{dy'}{dM_a}\right) \tag{95}$$

The C and H are in seconds of arc, just like the other four constants A, B, F, G. The V_r is thus given in seconds of arc per year. Dividing by the parallax gives the result in astronomical units per year or in 4.74 km/sec (because 1 astronomical unit per year = 149.5 million km/31.56 million seconds = 4.74 km/sec). For V_r expressed in kilometers per second we find thus:

$$V_r = \frac{dz}{dt} = L\frac{dx'}{dM_a} + N\frac{dy'}{dM_a} \tag{96}$$

where, if π_a is the absolute parallax:

$$\left.\begin{aligned} \pi_a L &= 4.74\, n\, C = 4.74\, n\, a \sin i \sin \omega \\ \pi_a N &= 4.74\, n\, H = 4.74\, n\, a \sin i \cos \omega \end{aligned}\right\} \tag{97}$$

The n is expressed here in degrees per year, while a is in seconds of arc. We are considering the radial velocity in the relative orbit and thus that of the fainter star with respect to the brighter one. This is quite different from the radial velocity curves in which one considers the absolute orbits with respect to the gravity center. If both velocity curves are known, the relative radial velocity can be found. This is what we are now considering. The L and N are proportional to a sin i, and therefore to the half amplitude K of the radial velocity curve, as we shall see later.

29. *Apparent orbit a straight line.* This is a particular case in which $i = 90°$ and $\Omega = \theta$. The best value of the position angle of the node is $\Omega = \bar{\theta}$. We can make here only a distance diagram. The position angle plotted against time gives a horizontal line. From the diagram we can determine P, for example from maximum to maximum. The T can be determined from the following consideration.

Periastron and apastron are $\frac{1}{2}P$ apart, while the distance to the median line are the same but opposite in sign with respect to this axis. The periastron point is situated on the steeper branch as the rate of change in distance is here largest. We can cut a rectangular slip of paper to a width equal to half that of the period and slide it along the curve until the edges, kept perpendicular to the median axis, cut equal ordinates on the curve.

One can also redraw the curve on tracing paper, turn it 180° and then advance the tracing by half a period. There are two pairs of intersections, but only one pair is separated by half the period. Apastron always lies on the more gradually sloping branch.

For the eccentricity we find $e = OS/Op = O_1S_1/O_1P_1$ (See Figure 36). We still have to find the a and ω.

$$\sin pSq = \sin v_q = \frac{qO}{qS} = \frac{b}{a} = \sqrt{(1 - e^2)} \qquad (98)$$

We find thus v_q and can find now M_q with help of tables published by F. Schlesinger and S. Udick. Because $M_q = (t - T)\, 2\pi/P$ we can now find the time interval necessary for the component to move from p to q. In other words q_3 can be found, and q_2 and q_1 follow. Further we have in similar triangles:

$$\frac{OO'}{b} = \frac{OO'}{a\sqrt{(1-e^2)}} = \frac{p_1O_1}{a}$$

$$OO' = p_1O_1 \sqrt{(1-e^2)}$$

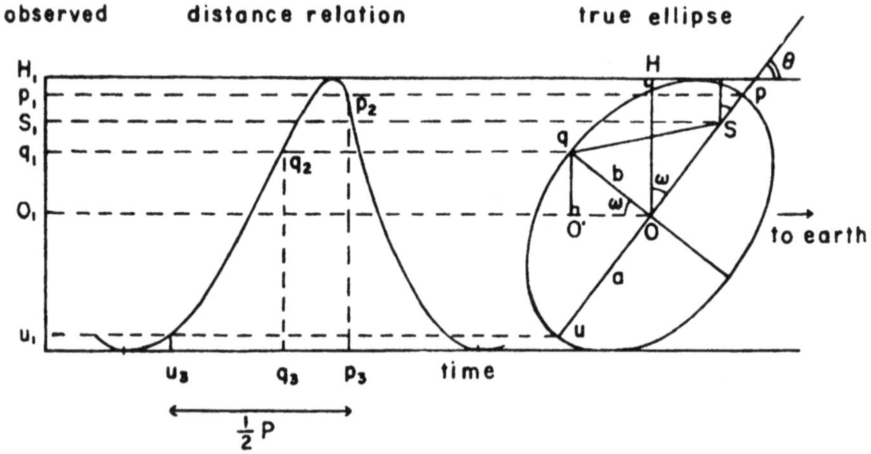

Figure 36. The apparent orbit of the visual double is observed as a straight line.

The ω is found as follows:

$$\tan \omega = \frac{qO'}{OO'} = \frac{q_1O_1}{OO'} = \frac{q_1O_1}{p_1O_1\sqrt{(1-e^2)}} \tag{99}$$

The a in seconds of arc is now determined:

$$\cos \omega = \frac{p_1O_1}{a}, \qquad a = \frac{p_1O_1}{\cos \omega} \tag{100}$$

There is still another method to find the last two elements.

$$O_1S_1 = OS \cos \omega = ae \cos \omega, \qquad \cos \omega = \frac{O_1S_1}{ae} \tag{101}$$

We will use now a property of the ellipse; the mathematical proof is given in chapter VI, paragraph 110.

$$\begin{aligned} OH^2 = O_1H_1^2 &= a^2\cos \omega + b^2\sin^2\omega \\ &= a^2 - a^2\sin^2\omega + b^2\sin^2\omega = a^2 - a^2e^2\sin^2\omega \\ &= a^2 - a^2e^2\left(1 - \frac{O_1S_1^2}{a^2e^2}\right) = a^2 - a^2e^2 + O_1S_1^2 \end{aligned}$$

Thus we find for the a:

$$a = \sqrt{\left(\frac{O_1H_1^2 - O_1S_1^2}{1 - e^2}\right)}. \tag{102}$$

We have to consider only the positive value. Substitution of this value of a in the formula for $\cos \omega$ gives the ω.

30. *Interferometer.* Until now we have considered visual double stars where we could really see two components, either visually or photographically. However, it is also possible to find out something about close pairs which cannot be seen as such. By visual means this is done with the interferometer. It consists of a metal plate with two equal and parallel openings, which is placed in the converging beam of light coming from the objective. We get now interference in the focal plane. The whole apparatus can rotate around the optical axis and the distance between the openings can be varied. In Figure 37 we place the

metal plate in front of the objective as was done in the older experiments, because the explanation is somewhat simpler. The distance between the openings is D.

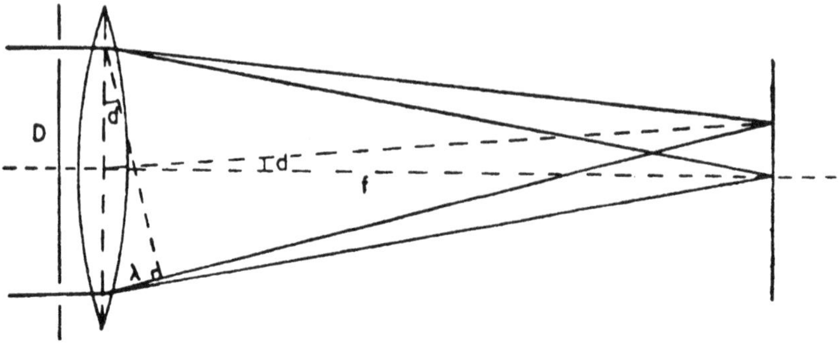

Figure 37. The interferometer consists of a plate with two equal and parallel openings.

For a single star the bright and dark interference lines are parallel. They are placed perpendicular to the line between the openings. Call λ the wavelength. The angular distance between two bright interference lines in the focal plane is $d = \lambda / D$. In case of a close double star one rotates the apparatus until the image in the focal plane looks the same as for a single star, and the interference lines cover each other as well as possible. The position angle of the line joining the openings is then the position angle θ of the double star. Now we vary the distance between the openings in such a way that the dark lines of one star coincide with the bright lines of the other component. Then there is a minimum visibility of the lines because the whole image looks like a normal star image (Figure 38). The duplicity can be found from the more or less complete disappearance of the dark fringes. The distance between the components is then:

$$\rho = \frac{d_0}{2} = \frac{\lambda}{2D_0} \tag{103}$$

It is found that the angular radius of the first dark ring surrounding the central disk of a usual image without interferometer is 1.22 λ/D, so that the interferometer increased the resolving power by a factor of 2.5 (Figure 39). It is possible to place the disk closer to the focal plane and so make the interferometer much simpler. Because we measure ρ and θ of the comparison star with respect to the principal star, we find the relative ellipse. The orbital determination is therefore the same as before.

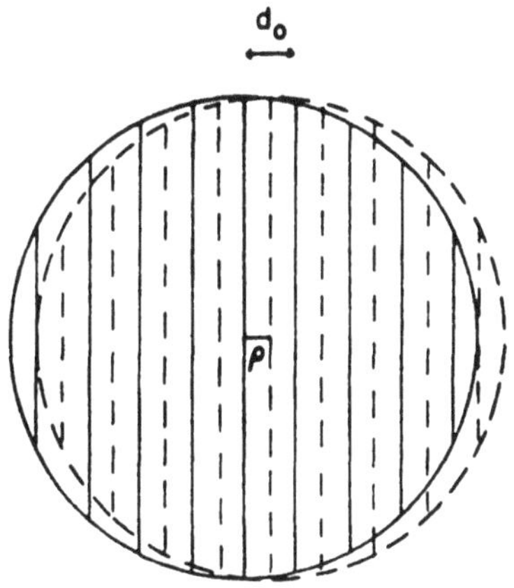

Figure 38. The distance between the openings of the interferometer is varied in such a way, that the dark lines of one star coincide with the bright lines of the other component.

The objective interferometer was also used at Mount Wilson Observatory to measure diameters of some super giants, which have the largest size. The D was made considerably larger than 100 inches with help of a system of mirrors, because the stellar diameters are still small in seconds of arc since the stars are so far away. The two halves of the disk of the star are considered separately as far as the interferometer is concerned.

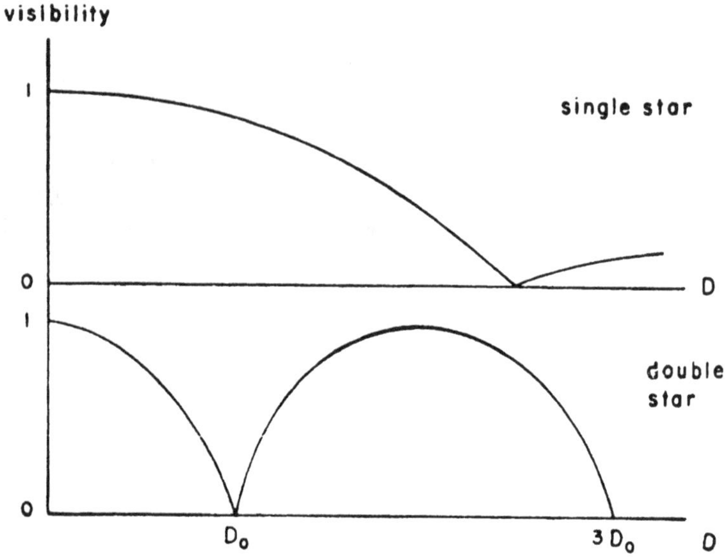

Figure 39. The relation between the visibility of the fringes and the separation of the interferometer apertures.

31. *Sum of the masses.* Double stars are almost the only objects in the universe whose masses can be determined directly. The reason is obvious. We cannot weigh a single star. But if we take two stars close together in space, then we can weigh one star or better determine its mass, if we can observe its effect on the other star. The force of gravity can be written as Newton's law but also as: Mass times acceleration. Division by the mass shows that the acceleration of one star depends only on the mass of the other star.

In visual double stars the distance between the components is observed in seconds of arc. Therefore, in the orbital computation the semi-major axis of the relative orbit is found in seconds of arc. In Kepler's third law $a = a_1 + a_2$ is expressed in astronomical units. Taking respectively a and 1 astronomical unit as base lines, we find:

$$a : 1 = a'' : \pi_a , \qquad a = \frac{a''}{\pi_a} \tag{104}$$

In case the absolute parallax is known we find for the visual double star:

$$\mathfrak{M}_1 + \mathfrak{M}_2 = \frac{a^3}{P^2} = \frac{a''^3}{\pi_a^3 P^2} \tag{105}$$

In this case we find thus the sum of the masses expressed in terms of the sun's mass. Note also that we find the semi-major axis of the relative ellipse or the sum of the semi-major axes of both absolute ellipses in astronomical units for the visual binary. To find the masses separately we need more information which we will discuss later in this chapter.

32. *Mass-luminosity relation.* A. S. Eddington found from theoretical considerations of stellar interiors a relation between absolute bolometric luminosity and the mass of a star. In practical work the relation derived is usually simplified and written as:

$$L = a\,\mathfrak{M}^b \tag{106}$$

He predicted $b = 3$ from his standard model and $b = 5.5$ from his more refined model. We make the drawing in logarithmic form where M is the absolute magnitude and $\mathfrak{M}$ is the mass (Figure 40).

$$M_{bol} = M = c - 2.5\,b \log \mathfrak{M} \tag{107}$$

For the components of a double star both following the relation we have then:

$$\Delta m_{bol} = \Delta M_{bol} = 2.5\,b \log \frac{\mathfrak{M}_1}{\mathfrak{M}_2} \tag{108}$$

A small correction must be applied, depending on the temperature of the star. Recent studies by P. van de Kamp, K. Aa. Strand and R. M. Petrie, where only the most accurate data are used, show that for solar-like stars one finds $b = 4.0$ from visual doubles, but for the luminous stars a somewhat smaller value from spectroscopic doubles. For solar-like stars with M_{bol} between $+7.5$ and 0.0 one finds in particular:

$$M_{bol} = 4.6 - 10.0 \log \mathfrak{M} \qquad (109)$$

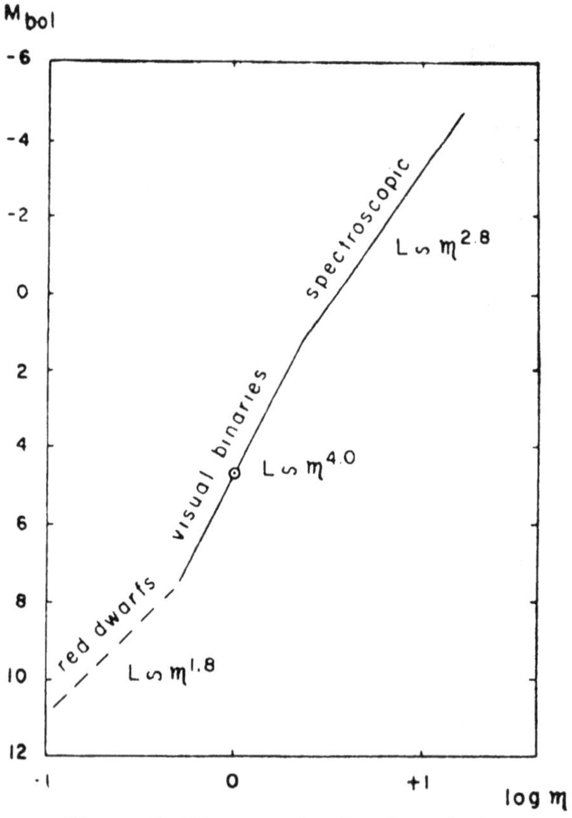

Figure 40. The mass-luminosity relation.

The red dwarfs give a well determined relation but with a different slope, namely $b = 1.8$. The white dwarfs do not follow any relation at all and are not considered here. There are also a number of serious deviations from the relation, the reality of which cannot be questioned, for example the components of W Ursae Majoris type variables.

33. *Dynamical parallax.* For a relative orbit we can find now from formula (105):

$$\pi_d = \frac{a''}{\sqrt[3]{P^2(\mathfrak{M}_1 + \mathfrak{M}_2)}} \tag{110}$$

The π_d is the dynamical parallax for visual double stars. If we start by taking $\mathfrak{M}_1 + \mathfrak{M}_2 = 2$, thus assuming both stars like the sun, then a provisional parallax follows from the formula, and the absolute magnitude from the well known relation (see chapter V, paragraph 71):

$$M = m + 5 + 5 \log \pi_a$$

Here M is the absolute magnitude and m the apparent magnitude. Under the assumption that the mass-luminosity relation holds, we can now find a better value $\mathfrak{M}_1 + \mathfrak{M}_2$. We can start now again, and by such a successive approximation after three or four times we find a constant value of the parallax which is the best possible π_d.

This method does not work for white dwarfs. For very distant binaries this method is a welcome addition. Note that the sum of the masses varies between rather small limits and that it enters into the formula under the cube root.

Another method is applied to a number of close visual binaries with known radial velocities. The orbital determination of the visual binary gives the angular size of the orbit. The relative radial velocities of the components give the linear size of the orbit. A comparison of the two gives the parallax.

34. *Astrometric double stars.* This is the case of a close double which is photographed as a single image on the photographic plate. One cannot measure the relative position of the faint component with respect to the brighter component and we cannot find the relative ellipse on the plate. However, we can measure the position with respect to comparison stars and find absolute ellipses in this way. At any time the positions of the components in the absolute ellipses are such that there is a mass balance with respect to the common gravity center, so that we have: mass times arm is constant. Let us suppose for a moment that we know the masses $\mathfrak{M}_1$ and $\mathfrak{M}_2$. We can then determine the gravity center at any moment in the relative ellipse. Then we know the distance of both components with respect to the gravity center G. In other words we can determine both absolute ellipses. The direction of motion is the same and the period and eccentricity are the same as in the relative orbit (Figure 41).

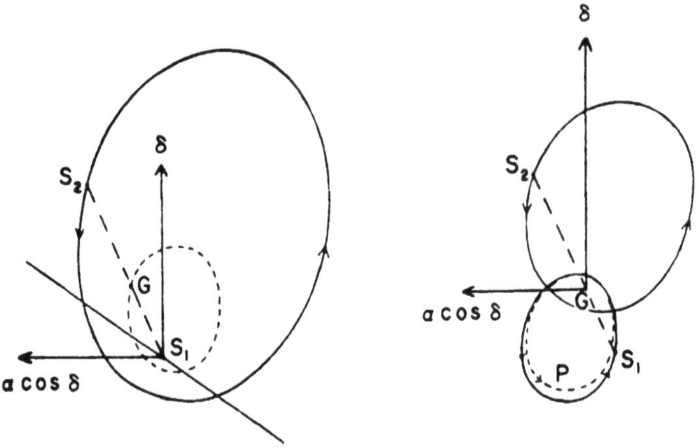

Figure 41. Left: The relative ellipse with the relative orbit of the gravity center G. Right: Both absolute ellipses together with the photocentric orbit.

This also holds for the true ellipses in the true plane, where we have both relative and absolute orbits. We have:

$$\mathfrak{M}_1 a_1 = \mathfrak{M}_2 a_2 , \qquad \frac{a_1}{a_2} = \frac{\mathfrak{M}_2}{\mathfrak{M}_1} \tag{111}$$

Here $a = a_1 + a_2$. We see also that the relative ellipse and the absolute ellipse are proportional. However, a positive factor for the fainter star and a negative factor for the brighter star are required.

35. *Resolved astrometric binary.* This is the case of a close double star observed as one image photographically but as two stars visually. Both apparent and true orbits are known. In other words from the visual double star the orbital elements are known, let us say by the Thiele-Innes method. Let us write for the four constants as in formula (81):

$$\begin{aligned} A &= a(), \qquad & F &= a() \\ B &= a(), \qquad & G &= a() \end{aligned}$$

For the apparent ellipse we can write then:

$$\left.\begin{aligned} y &= \delta &&= Ax' + Fy' = a\left(\frac{A}{a}x' + \frac{F}{a}y'\right) \\ x &= \alpha \cos \delta &&= Bx' + Gy' = a\left(\frac{B}{a}x' + \frac{G}{a}y'\right) \end{aligned}\right\}$$

We note that without the factor a in front of the parentheses we would have the unit ellipse. Now we consider first the negative unit ellipse, which is thus already in its correct orientation in the plane of the sky. The four constants become:

$$\left.\begin{aligned} (a) &= -\frac{A}{a}, \qquad & (f) &= -\frac{F}{a} \\ (b) &= -\frac{B}{a}, \qquad & (g) &= -\frac{G}{a} \end{aligned}\right\} \tag{112}$$

We multiplied by -1. The negative unit ellipse becomes:

$$\left.\begin{aligned} Q_\delta &= (a)\,x' + (f)\,y' \\ Q_\alpha &= (b)\,y' + (g)\,y' \end{aligned}\right\} \qquad (113)$$

These are called the orbital factors and are thus the coordinates of the negative unit ellipse in the plane of the sky.

36. *Photocentric orbit.* The equation of condition for the resolved binary is now:

$$\left.\begin{aligned} \xi &= c_x + \mu_x t + P_\alpha \pi_r + Q_\alpha\, \alpha \\ \eta &= c_y + \mu_y t + P_\delta \pi_r + Q_\delta\, \alpha \end{aligned}\right\} \qquad (114)$$

The known quantities are the measured coordinates ξ, η, the time t in years, the parallax factors P_α, P_δ, and the orbital factors Q_α, Q_δ. The unknown quantities are the zero point c_x, c_y, the proper motion μ_x, μ_y, the relative parallax π_r, and the semi-major axis α of the astrometric orbit. Because the period of the orbital motion is much longer than the parallactic motion period of one year, both can be separated in the least squares solution. Now we observe again a spiral but one of unequal spacing. The part of the orbital motion around periastron gives best results because the speed is there the largest and also the direction changes most rapidly (Figure 42).

If the intensity of the faint component is zero, as for a dark companion, we would have $\alpha = a_1$. It was found first historically in the cases of Sirius and Procyon that the proper motion was not linear. This was done visually by absolute methods. It was explained correctly as caused by companions of reasonable mass which were too faint to be seen. Much later they were discovered as very faint stars, namely the first white dwarfs. The parallactic ellipse was very small and not taken into account. From the mass balance, formula (111) we have:

$$\frac{\alpha}{a} = \frac{a_1}{a} = \frac{a_1}{a_1 + a_2} = \frac{\mathfrak{M}_2}{\mathfrak{M}_1 + \mathfrak{M}_2} = B_2 \qquad (115)$$

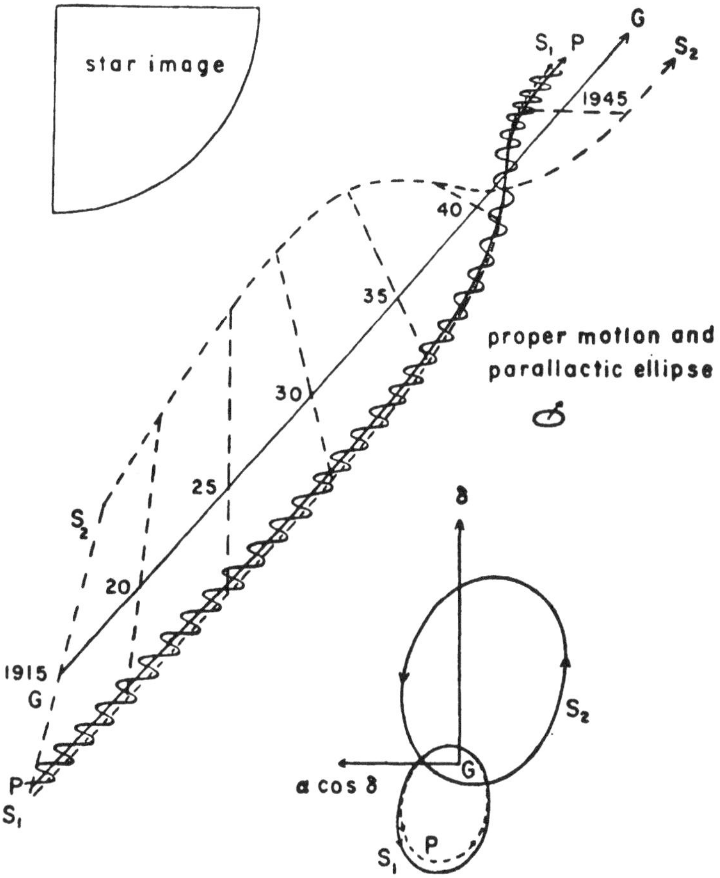

Figure 42. Observed photocentric spiral of an astrometric double (99 Her). See the size of the star image on the same scale.

Here B_2 is the fractional mass of the companion in terms of the total mass.

If the secondary component has a certain brightness, we measure the so-called photocentric orbit, corresponding with α. Given the intensities l_1 and l_2 we can see the situation as a light balance. The photocenter is the measured center of the combined image. If the distance between the two stars is taken as unity, then the distance β of the primary from the photocenter is:

$$\frac{\beta}{1} = \frac{l_2}{l_1 + l_2} \tag{116}$$

We can convert this into magnitude because we have:

$$m = -2.5 \log l, \qquad l = 10^{-0.4m}$$

The expression for β becomes then:

$$\beta = \frac{10^{-.4m_2}}{10^{-.4m_1} + 10^{-.4m}} = \frac{1}{1 + 10^{-0.4\Delta m}} \tag{117}$$

The Δm is the difference in magnitude between the components. The relation between β and Δm is given in Figure 43. This relation has been checked at Yerkes Observatory with artificial

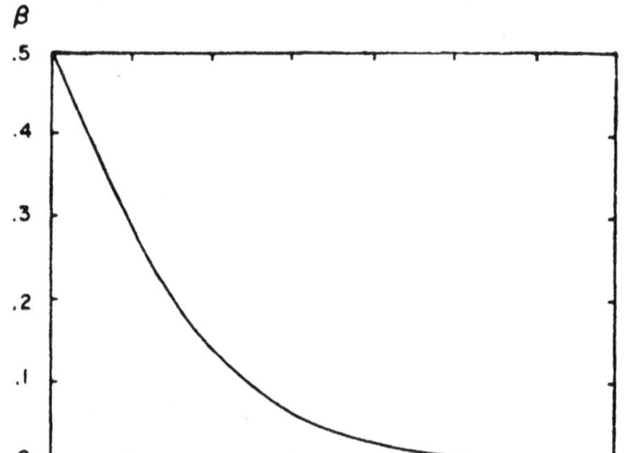

Figure 43. The relation between the distance β of the primary from the photocenter and the magnitude difference of the components.

stars. The fractional difference between barycenter or gravity center and the photocenter is therefore $B_2 - \beta$, and we get:

$$\frac{\alpha}{a} = B_2 - \beta, \qquad \alpha = (B_2 - \beta)a \tag{118}$$

The photocentric orbit is thus smaller than that of the bright component. For two stars with exactly the same masses and the same intensities we find $B_2 = \beta = 0.5$, thus $\alpha = 0$. Thus no orbital motion shows up in the photocentric orbit. But still this is a valuable result, if we know from other sources that it is a double star.

37. *Unresolved astrometric binary.* In the case where the astrometric double is unresolved, which means not observed visually or with the interferometer as a double, we do not obtain the orbital elements of the apparent ellipse. Sometimes it is a spectroscopic double, and we can get the elements from that, but otherwise we have no additional data. We then write our spiral as:

$$\left.\begin{aligned} \xi &= c_x + \mu_x t + P_\alpha \pi_r + (B)x' + (G)y' \\ \eta &= c_y + \mu_y t + P_\delta \pi_r + (A)x' + (F)y' \end{aligned}\right\} \tag{119}$$

We can find the dynamical elements P, T, e according to Zwiers' method, but we have then to determine the invisible barycenter. This may be difficult, and usually we follow a method by P. van de Kamp. Correct provisionally for the zero point, proper motion and parallax. We consider thus only the case:

$$\left.\begin{aligned} \Delta x &= \alpha Q_\alpha = \alpha\,(b)x' + \alpha(g)y' = (B)x' + (G)y' \\ \Delta y &= \alpha Q_\delta = \alpha\,(a)x' + \alpha(f)y' = (A)x' + (F)y' \end{aligned}\right\} \tag{120}$$

In these equations is $(B) = \alpha(b) = -\alpha\, B/a$, etc.

We can plot the αQ_α against time and the same for αQ_δ. See Figure 44. We have then two displacements curves. The P can be found from the repetition of the same displacement. The T can be found from the same considerations as was done with the case of a visual double star apparently moving in a straight line. We have to draw the median line. The e can now be found as follows. Call r_p, r_a the radius vectors to periastron and apastron respectively and V_p, V_a the velocities in the true orbit at these points respectively. Then by Kepler's second law it follows:

$$r_p V_p = r_a V_a, \qquad a(1-e)\, V_p = a(1+e)\, V_a$$

$$\frac{V_p}{V_a} = \frac{1+e}{1-e} \tag{121}$$

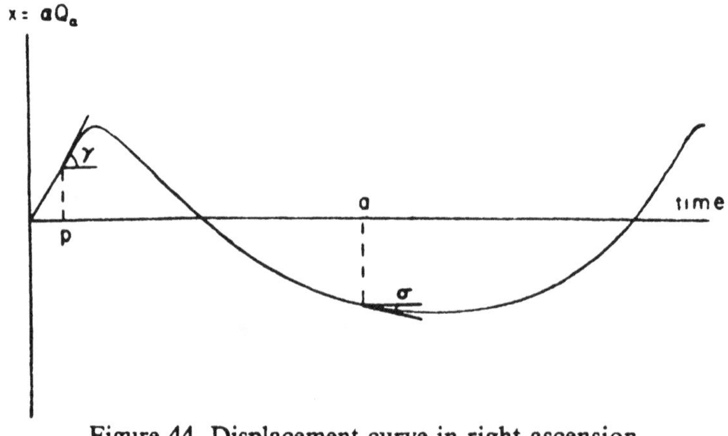

Figure 44. Displacement curve in right ascension.

The ratio of the velocities remains the same by projection in the plane of the sky, and also the same if we consider the components in the direction of the right ascension. We thus take the tangent of the slope at periastron and apastron. The negative sign indicates oppositely directed velocities.

$$\frac{\tan \gamma}{\tan \sigma} = -\frac{1+e}{1-e} \tag{122}$$

From the expression the e can be found. But then we are able to find the x', y' and solve the (A), (B), (F), (G) by least squares.

38. *Mass determination.* For the resolved astrometric binary we found:

$$\alpha = (B_2 - \beta)\, a, \qquad B_2 = \frac{\mathfrak{M}_2}{\mathfrak{M}_1 + \mathfrak{M}_2} = \frac{\alpha}{a} + \beta$$

B_2 is the fractional mass. We also find the other fractional mass B_1 because $B_1 + B_2 = 1$. Both are thus expressed in terms of the sum of the masses as a unit. Division gives the ratio of masses for the resolved astrometric binary.

$$\frac{\mathfrak{M}_2}{\mathfrak{M}_1} = \frac{a_1}{a_2} = \frac{B_2}{B_1} = \frac{B_2}{1 - B_2} \tag{123}$$

Because the binary is resolved we can find the β from the magnitude difference and thus compute this mass ratio.

Further we know the sum of the masses from the observations of the visual double star. The combination with the mass ratio gives then the masses separately, each thus expressed in the sun's mass as a unit. The dimensions of the system are in astronomical units which can be converted into kilometers.

For the unresolved binary we do not know the β because the Δm is unknown. We can only find the mass function.

$$\mathfrak{M}_1 + \mathfrak{M}_2 = \frac{\alpha^3}{(B_2 - \beta)^3\, \pi_a^3\, P^2}$$

$$(B_2 - \beta)^3\, (\mathfrak{M}_1 + \mathfrak{M}_2) = \frac{\alpha^3}{\pi_a^3\, P^2} \tag{124}$$

It is found that the masses of the stars are not very different, and most of them range between 20 and 0.5 times the solar mass. Massive stars are also more luminous.

39. *Résumé.* We have now considered the following cases:

TABLE I. SUMMARY OF METHODS

Case	*Visual*	*Method*	*Photo-visual*	*Method*
wide pair	two stars	micrometer	two images	sequences
close pair	two stars	micrometer	one image	comp. stars, resolved binary
very close pair	one star	interfero-meter	one image	comp. stars, unresolved binary
ellipse	relative orbit		absolute orbit	

The orbital elements can be derived but the sign of the inclination becomes known only when spectroscopic observations are available. The relative orbit gives us the sum of the masses of the components. The absolute orbit gives us the mass ratio and for the resolved astrometric binary the combination of both gives us the masses separately.

Some components of visual double stars are themselves binaries and give a disturbed ellipse. Here one determines the Kepler ellipse first and eliminates this orbit. The residuals can then be studied for analyzing this perturbation.

40. *Proofs of certain formulae.* We will give here the mathematical proofs of some formulae used in the previous paragraphs.

(1) In the rectangular spherical triangle we have (Figure 31):

$$\cos i = \frac{\tan(\theta - \Omega)}{\tan(v + \omega)}$$

$$\tan(v + \omega) = \tan(\theta - \Omega)\sec i$$

(2) According to a property of the ellipse we have (Figure 45):

$$S_2K : LK = b : a$$

$$r \sin v : a \sin E = b : a, \qquad r \sin v = b \sin E$$

$$S_1K = r \cos v = a \cos E - ae, \qquad r \cos v = a(\cos E - e)$$

Squaring and adding gives, if we substitute $b^2 = a^2(1 - e^2)$:

$$\begin{aligned} r^2 &= b^2 \sin^2 E + a^2 \cos^2 E + a^2e^2 - 2a^2e \cos E \\ &= a^2\sin^2 E - a^2e^2\sin^2 E + a^2\cos^2 E + a^2e^2 - 2a^2e \cos E \\ &= a^2(1 + e^2 - e^2\sin^2 E - 2e \cos E) \\ &= a^2(1 + e^2\cos^2 E - 2e \cos E) = a^2(1 - e \cos E)^2 \\ r &= a(1 - e \cos E) \end{aligned}$$

Further, $\cos v = 1 - 2 \sin^2 \frac{v}{2}$ and we have:

$$\begin{aligned} 2r \sin^2 \frac{v}{2} &= r - r \cos v = a(1 - e \cos E) - a(\cos E - e) \\ &= a(1 - e \cos E - \cos E + e) \\ &= a(1 + e)(1 - \cos E) \end{aligned}$$

In a similar way we find:

$$2r \cos^2 \frac{v}{2} = a(1 - e)(1 + \cos E)$$

By division we find:

$$\tan^2 \frac{v}{2} = \frac{1 + e}{1 - e} \times \frac{1 - \cos E}{1 + \cos E} = \frac{1 + e}{1 - e} \tan^2 \frac{E}{2}$$

$$\tan \frac{v}{2} = \sqrt{\frac{1 + e}{1 - e}} \tan \frac{E}{2}$$

(3) According to Kepler's second law we have:

$$\frac{S_1S_2A}{\pi ab} = \frac{t - T}{P}$$

$$\text{area } S_1S_2A = \pi ab \frac{t - T}{P} = \tfrac{1}{2}\, ab\, n(t - T) = \tfrac{1}{2}\, ab\, M_a$$

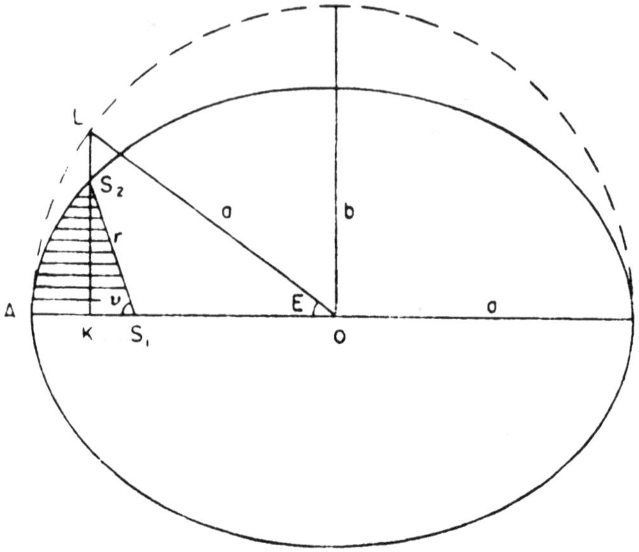

Figure 45. The Kepler equation.

We compute now this area in another way as the sum of two areas.

$$\begin{aligned}
\text{area } S_1S_2K &= \tfrac{1}{2}\, S_1K \times S_2K = \tfrac{1}{2}\, a(\cos E - e)\frac{b}{a}\, a \sin E \\
&= \tfrac{1}{2}\, ab \sin E\,(\cos E - e) \\
\text{area } S_2KA &= \frac{b}{a}\,\text{area LKA} = \frac{b}{a}\left(\tfrac{1}{2}\, a^2 E - \tfrac{1}{2}\, a^2 \sin E \cos E\right) \\
&= \tfrac{1}{2}\, ab\,(E - \sin E \cos E) \\
\text{area } S_1S_2A &= \tfrac{1}{2}\, ab\, M_a = \tfrac{1}{2}\, ab\,(E - e \sin E)
\end{aligned}$$

From this we find the Kepler equation.

$$M_a = E - e \sin E = \frac{2\pi}{P}\,(t - T) = n\,(t - T)$$

(4) The axes of the ellipse are $a = \alpha$, $b = a \cos i = \beta$. The conjugate diameters have the semi-major axes a_1 and b_1. The

angle between a and a_1 is λ, between a and b_1 is ν, between a_1 and b_1 is σ. We have then in Figure 46:

$$\begin{cases} x_0 = +\, a\cos E = a_1 \cos\lambda, & y_0 = b\sin E = a_1 \sin\lambda \\ x_1 = -\, a\sin E = b_1\cos\nu, & y_1 = b\cos E = b_1\sin\nu \end{cases}$$

By squaring and adding we find:

$$c^2 = a^2 + b^2 = a_1^2 + b_1^2$$

Further $\sigma = \nu - \lambda$, $\quad \sin\sigma = \sin\nu\cos\lambda - \cos\nu\sin\lambda$

$$a_1b_1\sin\sigma = b\cos E \times a\cos E + a\sin E \times b\sin E$$

$$a_1b_1\sin\sigma = ab, \qquad 4a_1b_1\sin\sigma = 4ab$$

These are the two theorems of Apollonius. The sum of the squares of two conjugate diameters is constant and equal to the sum of the squares of the axes. The area between the parallelogram on two conjugate diameters is constant and equal to the rectangle on the axes.

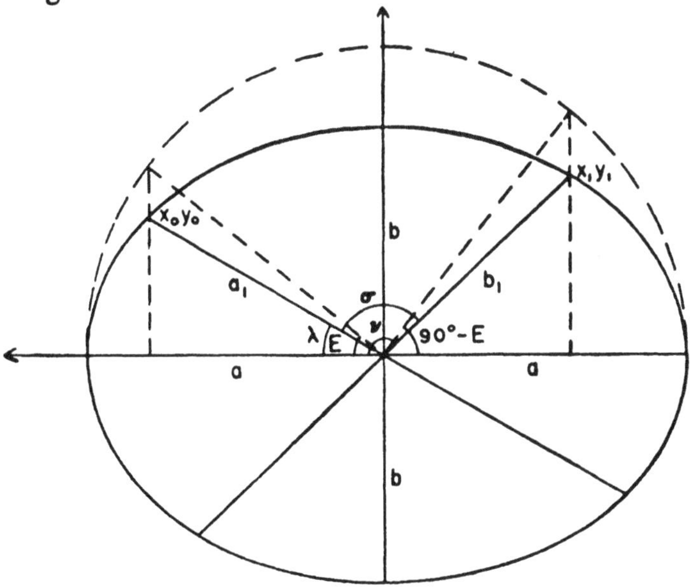

Figure 46. The theorems of Apollonius.

REFERENCES

METHOD OF KOWALSKY AND GLASENAPP

M. Kowalsky: *Proc. Kasan Imp. Un.*, 1873.
S. Glasenapp: *M.N.*, **49**, 276, 1889.
R. G. Aitken: *The binary stars*. McGraw-Hill Book Co., New York and London, 80, 1935.

METHOD OF ZWIERS AND RUSSELL

H. J. Zwiers: *A.N.*, **139**, 369, 1895.
H. N. Russell: *A. J.*, **19**, 9, 1898.
G. C. Comstock: *A.J.*, **31**, 33, 1918.

METHOD OF THIELE AND INNES

T. N. Thiele: *A.N.*, **104**, 245, 1883.
W. H. van de Bos: *Union Circ.*, No. 68, 354, 1926; No. 86, 261, 1932.
Tables for elliptic rectangular coordinates: *Union Circ.*, No. 71, 391, 1926.
K. Aa. Strand: *Leiden Ann.*, **18**, part 2, 64, 1937.

APPARENT ORBIT A STRAIGHT LINE

F. Henroteau: *P.A.S.P.*, **29**, 195, 1917.
K. Laves: *A.J.*, **37**, 97, 1926.
R. T. Crawford: *L.O.B.*, **14**, 6, 1928.

INTERFEROMETER

A. A. Michelson: *Phil. Mag.*, **30**, 1, 1890.
J. A. Anderson: *Ap. J.*, **51**, 263, 1920; *Mt. W. Contr.* No. 185.
H. Spencer Jones: *M.N.*, **82**, 513, 1922.
W. S. Finsen: *Pop. Astr.*, **59**, 399, 1951.

ASTROMETRIC DOUBLE STARS

P. van de Kamp: *Pop. Astr.*, **52**, 53, 1944.
P. van de Kamp: *A.J.*, **52**, 185 and 189, 1947.
P. van de Kamp: *Encyclopedia of Physics*, **50**, 187, 1958.

MASS-LUMINOSITY RELATION

G. P. Kuiper: *Ap. J.*, **88**, 472, 1938.
H. N. Russell and C. E. Moore: *The masses of the stars*, Un. of Chicago Press, Chicago, 1940.
K. Aa. Strand and R. G. Hall: *Ap. J.*, **120**, 322, 1954.
P. van de Kamp: *A.J.*, **59**, 447, 1954.
K. Aa. Strand: *R.A.S.C. Journal*, **51**, 46, 1957; *Contr. Dearborn Obs.* No. 10.

CATALOGUES OF ORBITAL ELEMENTS

W. S. Finsen: *Union Circ.*, No. 91, 23, 1934.
R. G. Aitken: *The binary stars*. McGraw-Hill Book Co., New York and London, 284, 1935.
B. Ekenberg: *Medd. Lund. Obs.*, Ser. 2, No. 94, 1938.
P. Muller: *Journal des Observateurs*, **36**, 61, 1953; **37**, 153, 1954.
P. Muller: *Circulaire d'Information*. Obs. Strasbourg et Meudon.
O. J. Eggen: *A.J.*, **61**, 405, 1956.

III

Spectroscopy

41. *The Spectrograph.* A GLASS PRISM SEPARATES WHITE LIGHT into its component wavelengths. We get the sharpest definition when the rays pass through the prism at the so-called minimum deviation, that is the least possible deviation from the original direction. This is also the position of maximum light transmission. Inside the prism the ray then makes equal angles with the faces of the prism. When γ is the angle of the prism, δ_m the angle of minimum deviation, and n_λ the refraction index, we have in Figure 47:

$$\sin \tfrac{1}{2}(\gamma + \delta_m) = n_\lambda \sin \tfrac{1}{2}\gamma \tag{125}$$

This holds for any angle γ but rigorously only for one wavelength.

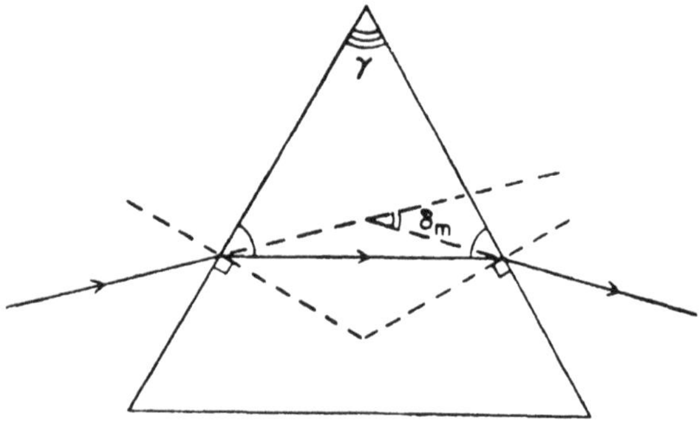

Figure 47. Minimum deviation of a monochromatic light ray in a glass prism.

A slit spectrograph usually contains 60° prisms. In the case of an objective prism where the angle γ is small, we have:

$$\frac{\sin \frac{1}{2}(\gamma + \delta_m)}{\sin \frac{1}{2}\gamma} = \frac{\frac{1}{2}(\gamma + \delta_m)}{\frac{1}{2}\gamma} = \frac{\gamma + \delta_m}{\gamma} = n_\lambda$$

$$\delta_m = (n_\lambda - 1)\gamma \qquad (126)$$

The spectroscope consists of a collimator, prism and refracting telescope arranged as in Figure 48. The slit is in the focal plane of the lens of the collimator so that a parallel beam reaches the prism. The outcoming separate beams of each different wavelength of light are brought into focus by the objective lens of the small telescope. In the focal plane we can place a photographic plate or view the spectrum with an eyepiece.

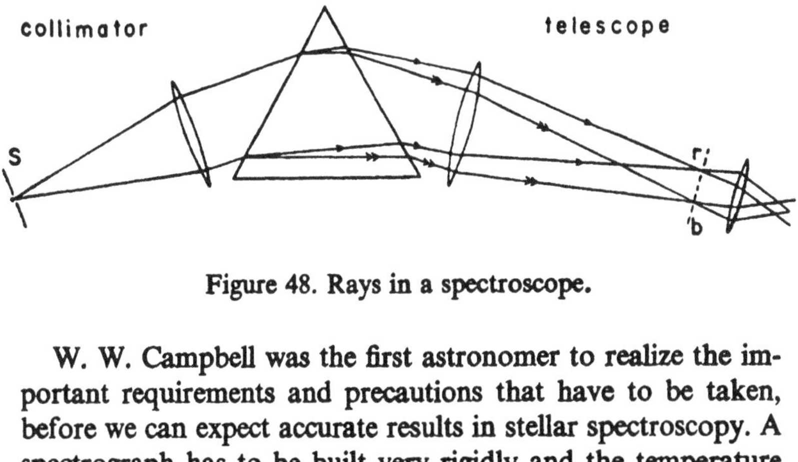

Figure 48. Rays in a spectroscope.

W. W. Campbell was the first astronomer to realize the important requirements and precautions that have to be taken, before we can expect accurate results in stellar spectroscopy. A spectrograph has to be built very rigidly and the temperature during a photographic exposure must be controlled and kept within very narrow limits (Plate VI). Care has to be taken so that as little light as possible is lost. The spectrum is an array of monochromatic slit images. If the slit is straight, these spectral lines are slightly curved by an amount depending upon the optical constants of the instrument and the wavelength of the line. The slit is often given a counter curve so that the absorption lines appear as straight lines for a short range in the spectrum.

Gaseous nebulae show a bright line spectrum. For the sun and stars we get an absorption line spectrum, where the continuous array of bright emission is interrupted by dark absorption lines. Above and below this spectrum we photograph a comparison spectrum of a source at rest with respect to the observer having many lines whose wavelengths have been accurately measured in the laboratory. An iron arc or spark spectrum, for example, gives many lines and is commonly used. The comparison source must be adjusted in such a way that its light follows very nearly the same path as the star light in the spectrograph. The instrument must be checked regularly with help of a planet like Venus for which the relative speed can be computed from the known motions in the solar system.

42. *Observation and measurement.* The slit must be in the focal plane of the telescope. Usually the slit plate is reflecting and inclined at a small angle to the collimation axis, so that the observer can guide the star directly on the slit. Constant and careful guiding is important. For a refracting telescope there is a slight dependence of the focal length upon temperature; for a reflecting telescope there is a larger dependence, and the focal point must be checked often during the night. We place the slit along a parallel of declination, thus east-west, to permit trailing of the image along the slit. We want the spectrum somewhat broadened, and therefore we move the star image slowly in the direction of the slit at a speed depending on the brightness of the star. In the case of a long exposure, we allow the star to trail the length of the slit many times. The clock drive is set fast or slow to permit the trailing. Because we want reasonable exposure times the slit width is sometimes increased for faint stars, at the expense of the resolution.

For the measurement of a spectrogram one follows the usual procedure already described in the first chapter. First adjustments of focus and alignment are made. Great care should be taken so that the illumination of the field is uniform. Beginning with a comparison line the settings are made continuously along the plate on good star and comparison lines as they chance to occur. The plate is then reversed 180° and the settings are repeated. The average value is then taken to eliminate several sources of systematic errors. The effects of accidental errors are minimized by employing a number of lines. For each spectral type a list of unblended lines can be made which are best suited for the purpose.

43. *Reduction.* First we identify the lines in the comparison spectra and find their wavelengths in a catalogue. For example, these standard wavelengths are given in the *Transactions* of the International Astronomical Union. These are measured in the laboratory with a high dispersion spectrograph employing a grating, on which thousands of parallel lines per mm are engraved. For such a grating the relation between the screw readings on the plate and the wavelength is almost linear.

Most stellar spectrographs employ prisms, though with very large telescopes grating spectrographs are now used extensively. In the case of a prism, we have on the photographic plate a deviation from the linear relation between the measured position or screw reading x and the wavelength λ. For a prismatic spectrum it is found that the relation can be closely expressed by an equilateral hyperbola of the form $(\Delta x)(\Delta\lambda) = H$, which is usually written as (Figure 49):

$$x - x_0 = \frac{H}{\lambda - \lambda_0} \tag{127}$$

This expression is known as the Hartmann formula. The constants x_0, λ_0, H have to be determined. Since it is not practicable to plot the dispersion curve, we use this formula in practice. We thus have the laboratory λ and measured x for three comparison lines and we get three Hartmann equations which are enough to solve x_0, λ_0, H algebraically. We can use now these constants for each measured comparison line and for the given

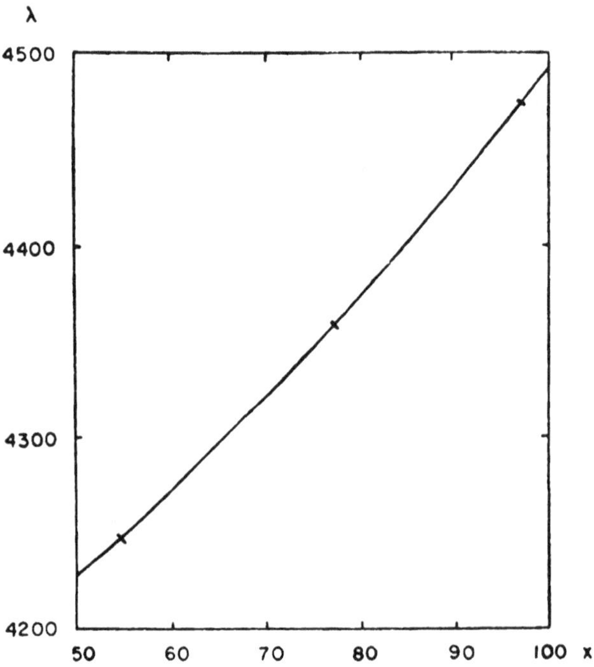

Figure 49. The Hartmann curve for the reduction of the spectral plate.

laboratory λ compute the corresponding x_c. We plot now the difference between the computed x_c and the measured x_m against x, which shows us the deviation of the real curve from the hyperbola through the three points. The measured star line position is now first corrected for this deviation. The corrected value for the star line x_s gives by substitution in the formula the computed wavelength for the star line. Finally, we have found now $\Delta\lambda$ = computed λ − laboratory λ, due to radial velocity of the star, or some other cause. We can divide the plate into various regions and do the same for each.

The Doppler formula now gives us:

$$\frac{V_m}{c} = \frac{\Delta\lambda}{\lambda} \tag{128}$$

$$V_m = 299{,}776\frac{\Delta\lambda}{\lambda} \quad \text{km/sec} \tag{129}$$

Here V_m is the measured radial velocity and c the velocity of light both measured in the same unit, usually km/sec. The radial velocity is a measure of relative speed in the line of sight and is independent of distance. It is the change of distance per time unit between the source and the observer, thus $V_m = dz/dt$. The V_m is thus positive when the distance increases.

If there are no complications due to shell spectra, gaseous currents, pulsation, and so on, we will find a constant radial velocity within the probable error for the different lines used on one given plate. The shift in wavelength is now found to be proportional to the wavelength so that an absorption line in the red shows a much larger shift than a line in the blue. Moreover, we have seen that this shift in wavelength is not proportional to the shift in position which is what we measure on the plate.

The dispersion or reduction curves for different plates vary only slightly for different temperatures. One can therefore use also a standard reduction curve derived for example from a solar spectrum, and compile this into a standard table. The dispersion at a certain wavelength is the number of Angstroms per mm. For another plate one studies the differences between the measurements and this table. Because both dispersion curves differ only slightly, these differences can be corrected graphically. These corrections are not always linear. By interpolation one finds the differences for the star lines.

A spectro-comparator is a special measuring microscope in which a spectral plate can be referred directly to a standard plate. This standard plate, usually of the sun, is taken with the same stellar spectrograph. On one of the carriages the star plate is placed and on the other movable one the standard plate. The microscope has two objectives which are so arranged that, by means of reflection prisms, corresponding parts of the two plates can be brought into focus in the same plane in one eyepiece. A silvered strip on the surface of one prism cuts out the central part of the solar spectrum and throws the star spectrum into its place. The same can be done with the comparison spectra of the two plates. It is possible to equalize the scale of the two plates by changing the relative magnifying powers of the two objectives.

After proper alignment of the spectral plates the corresponding comparison lines of solar and stellar spectrograms are brought in the same straight line. A setting is then made with the corresponding lines of the solar and stellar spectra in the same straight line. The difference between the micrometer readings in the two positions is the displacement of the star lines relative to the solar lines. The settings are also made with the plates in reversed position. Usually the plate is divided in several regions. After applying the correction for the known velocity of the solar plate, we find the radial velocity of the star relative to the observer.

While absolute wavelengths are not important for radial velocities the relative wavelengths of the comparison lines have to be known accurately. An error of $\pm$ 0.01 A produces an error of about 1 kilometer per second.

44. *Daily rotation.* The measured radial velocity consists of a sum of different radial velocities because the change of distance between the star and the observer is the combined effect of several different causes. First, because of the daily rotation of the earth each observer is constantly moving towards the east. We will call this V_d and apply a correction $-V_d$ to reduce the measured radial velocity to the center of the earth. Further the observer is revolving with the earth around the sun. This gives a yearly component V_y. If we apply the correction $-V_y$ then the yearly radial velocity is reduced to the sun and made heliocentric. The radial velocity of a star given in publications or catalogues is this heliocentric radial velocity V_r freed from the daily rotation and the yearly revolution. By this reduction two observatories observing the same star at the same universal time should find the same radial velocity within the probable errors, if no systematic effects are present.

$$V_m = V_d + V_y + V_r$$
$$V_r = V_m - V_d - V_y \tag{130}$$

The radius of the earth is 6378 km; the number of mean solar seconds in a sidereal day is 86,164 sec. The equatorial speed due to rotation is therefore 0.47 km/sec, and the speed for an observer at latitude φ is 0.47 cos φ. Let S be the place on earth where star S is in the zenith (Figure 50). We must then project this speed on the line ES to get the influence on the radial velocity. In Δ SGD we have:

$$\cos \mathrm{DS} = \cos \delta \cos (90 + \tau) = -\cos \delta \sin \tau$$

where τ = hour angle, δ = declination of star. In the figure we draw the speed of the observer and move it parallel to the center of the earth. The projection factor is therefore:

$$\cos (180 - \mathrm{DS}) = \cos \delta \sin \tau$$

The projection of the speed on the line is positive in the figure

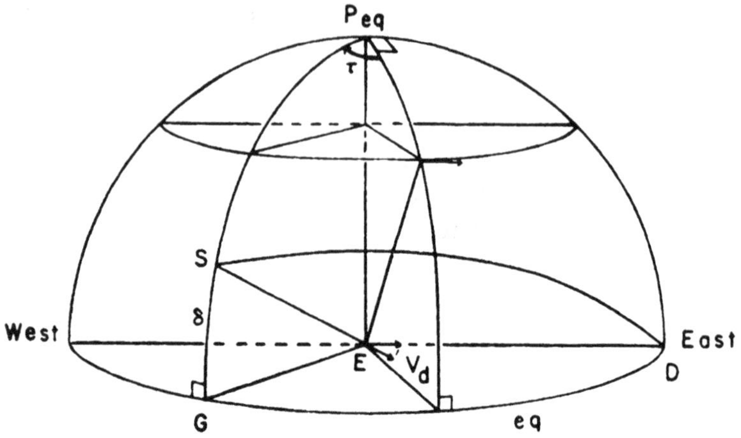

Figure 50. The spherical triangle SDG.

because the distance ES increases as far as the motion of the daily rotation is concerned. Multiplication by the speed of the observer now gives V_d. The correction is thus:

$$- V_d = - 0.47 \cos \varphi \cos \delta \sin \tau$$
$$- V_d = A \sin \tau, \quad A = - 0.47 \cos \varphi \cos \delta \qquad (131)$$

For one observatory φ = constant. For the star under observation δ = constant. In the meridian the correction is zero because we then move perpendicular to the line of sight. The sin τ is positive when the star is west of the local meridian.

45. *Yearly revolution.* We consider first the case of the circular orbit of the earth around the sun. We have: 1 astronomical unit = 149.5 million kilometers; one year has t = 31.56 million seconds. Thus the speed of the earth in its orbit is $V_c = 2\pi\ a/t = 2\pi \times 4.74 = 29.76$ km/sec. The direction of this velocity is toward a point in the ecliptic 90° behind the sun, since the velocity and radius vectors are mutually perpendicular. Thus the longitude of F is (☉ − 90°). The star S has λ and β as ecliptic

coordinates. Thus $\angle$ GEF $= \lambda - (\odot - 90°) = \lambda - \odot + 90°$ (Figure 51). In Δ SFG we have according to the cosine rule:

$$\cos \mathrm{SF} = \cos\beta \cos(\lambda - \odot + 90) = \cos\beta \sin(\odot - \lambda)$$

The projection V_c cos SF is directed towards the star S thus decreasing the distance between E and S. By definition the radial velocity is then negative. In this figure cos SF is positive.

The correction in the radial velocity caused by the yearly revolution is therefore $-V_y = V_c \cos \mathrm{SF}$.

$$-V_y = 29.76 \cos\beta \sin(\odot - \lambda)$$

$$-V_y = C \sin(\odot - \lambda), \qquad C = 29.76 \cos\beta \qquad (132)$$

If we apply this correction, the observation is reduced to the center of the sun and thus made heliocentric.

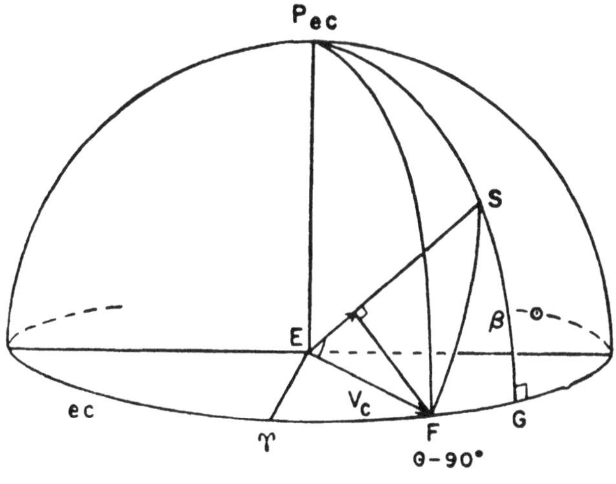

Figure 51. The spherical triangle P_{ec}SF.

We now consider the case of the earth's motion in its ellipse of eccentricity $e = 0.01674$. The velocity V in the ecliptic plane has the direction of the tangent to the ellipse at E. This velocity has components V_1 and V_2, where V_1 is perpendicular to the radius vector and V_2 is perpendicular to the major axis. A theorem of the two body problem proves that they are constant all the time. V_2 keeps the same size and direction; V_1 keeps the same size but changes its direction as the radius vector sweeps around (Figure 52). We have:

$$h = \frac{2\pi ab}{t}, \qquad p = \frac{b^2}{a} = a(1 - e^2)$$

For the components we find then:

$$\left.\begin{aligned} V_1 &= \frac{h}{p} = \frac{2\pi a}{t}\frac{a}{b} = \frac{2\pi a}{t}\frac{1}{\sqrt{(1-e^2)}} = 1.00014\ V_c = 29.76 \\ V_2 &= e\frac{h}{p} = 0.01674 \times 1.00014\ V_c = 0.01674\ V_c = \ \ 0.50 \end{aligned}\right\}$$

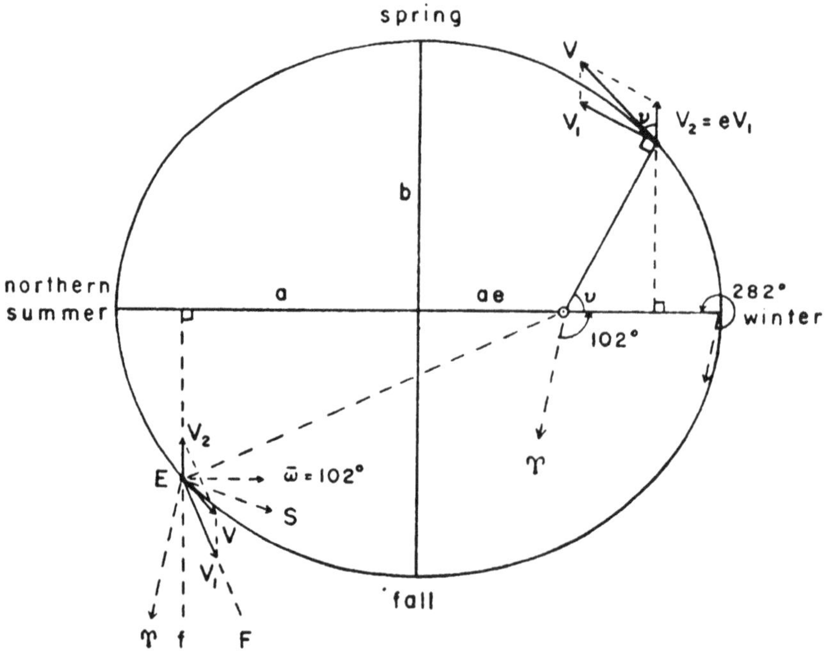

Figure 52. The velocity components in the elliptical orbit.

At perihelion the velocity will be $(1 + e)V_c$ or eV_c larger than for circular motion. This adds 0.50 km/sec and already gives us an idea about the coefficient of the additional term caused by the eccentricity of the earth's orbit.

In Figure 53 the V_1 is perpendicular to the earth-sun direction. Δ SFG is the same as before, and after the projection we will find thus an expression similar to the previous one. We have only to replace V_c by V_1. For V_2 we will use Δ SfG. We have now: $\angle \text{GEf} = \lambda - \bar{\omega} + 90°$ where $\bar{\omega}$ is the perihelion longitude of the earth. This is similar to the first expressions for V_1 and

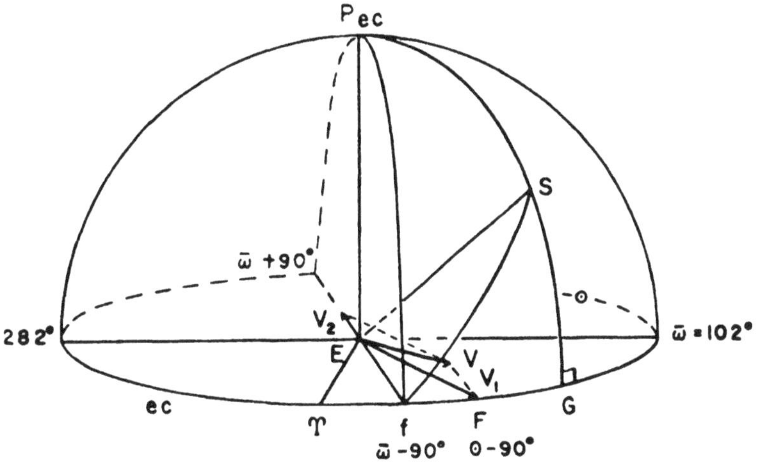

Figure 53. The spherical triangle P_{ec}Sf.

V_2, but we have to replace ⊙ by $\bar{\omega}$. The V_2 is also in the direction opposite to Ef, so that we have to introduce a negative sign. The correction in the radial velocity can now be written without further computation.

$$\left.\begin{aligned} -V_{y1} &= 29.76 \cos\beta \sin(\odot - \lambda) \\ -V_{y2} &= -0.50 \cos\beta \sin(\bar{\omega} - \lambda) \end{aligned}\right\} \qquad (133)$$

For the earth $\bar{\omega} = 102°$ as seen from the sun. This value changes slowly with time because the major axis of the earth's orbit is moving. The correct value can be found in the almanac; for all practical purposes it can be taken as a constant. For the sun as seen from the earth we thus find 282°. In other words the sun is in perigee during one of the first days of January. It is seen that the V_{y2} has to be taken into account. The total correction is:

$$-V_y = \{29.76 \sin(\odot - \lambda) - 0.50 \sin(\bar{\omega} - \lambda)\} \cos\beta \qquad (134)$$

For an ellipse the sun's longitude $\odot$ is not exactly proportional to time. If we plot the correction against time, we will find a somewhat distorted sine curve which moreover is shifted in the ordinate direction (Figure 54).

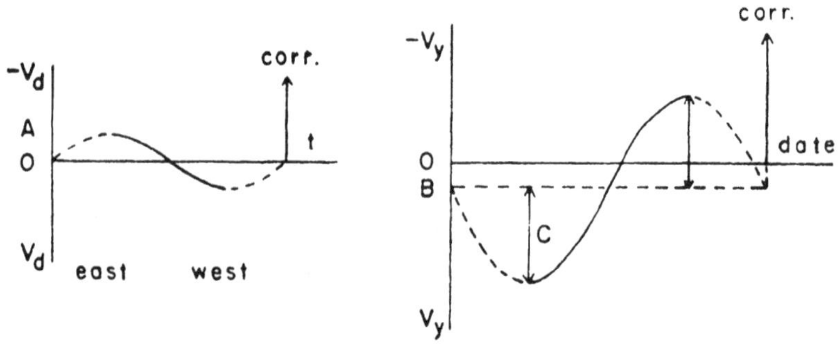

Figure 54. The daily and yearly corrections.

Summarizing, we have thus the following corrections.

$$\left.\begin{aligned} -V_d &= A \sin\tau \\ -V_y &= B + C\sin(\odot - \lambda) \end{aligned}\right\}$$

where the constants have the following values:

$$\left.\begin{aligned} A &= -\ \ 0.47 \cos\varphi \cos\delta \\ B &= -\ \ 0.50 \cos\beta \sin(\bar{\omega} - \lambda) \\ C &= +29.76 \cos\beta \end{aligned}\right\}$$

If spectra of the same star are taken at the same sidereal time but on different dates, one expects to find:

$$V_r'' - V_r' = \frac{2\pi a}{t} \frac{1}{\sqrt{(1-e^2)}} \left\{ \sin(\odot' - \lambda) - \sin(\odot'' - \lambda) \right\} \cos\beta \tag{135}$$

From this one can determine a and thus the solar parallax. The largest value of $V_r'' - V_r'$, and thus the most accurate value of a, is obtained when $\beta = 0$ and $\odot' - \lambda = -(\odot'' - \lambda) = 90°$ or 270°. The star must be near the ecliptic and the two sets of observations must be made six months apart, when the longitude difference between the sun and the star is $\pm$ 90°. For a circular orbit we get the factor V_c before the braces. For $\beta = 0$ at the most favorable positions we have:

$$V_r'' - V_r' = \frac{2\pi a}{t} 2 = 2\,V_c \tag{136}$$

The radial velocity of the star itself is eliminated here (see Figure 55).

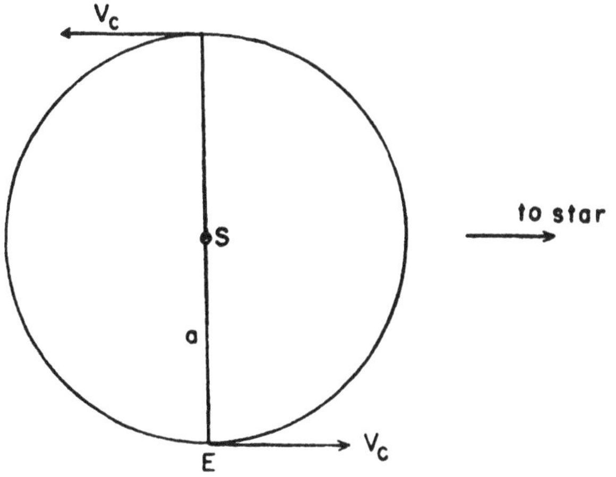

Figure 55. Determination of the solar parallax from radial velocity observations.

46. *Heliocentric correction in* α, δ. It may be convenient to have the correction when the right ascension and declination of the star are known. We will first consider the case of a circular orbit. Call the coordinates of F (A_1, D_1). In Δ $P_{eq}SF$ we have according to the cosine rule (Figure 56).

$$\begin{aligned}\cos SF &= \sin D_1 \sin \delta + \cos D_1 \cos \delta \cos (\alpha - A_1) \\ &= \sin D_1 \sin \delta + \cos D_1 \cos \delta \cos \alpha \cos A_1 \\ &\quad + \cos D_1 \cos \delta \sin \alpha \sin A_1\end{aligned}$$

In Δ ♈ FT we have according to the three rules:

$$\left.\begin{aligned}\sin \odot &= \cos D_1 \cos A_1 \\ -\cos \odot \sin \epsilon &= \sin D_1 \\ -\cos \odot \cos \epsilon &= \cos D_1 \sin A_1\end{aligned}\right\}$$

We therefore find by substitution:

$$\cos SF = \cos \alpha \cos \delta \sin \odot - (\underline{\sin \epsilon \sin \delta} + \underline{\cos \epsilon} \sin \alpha \cos \delta) \times \cos \odot$$

The yearly correction $-V_y$ is found by multiplying by V_c.

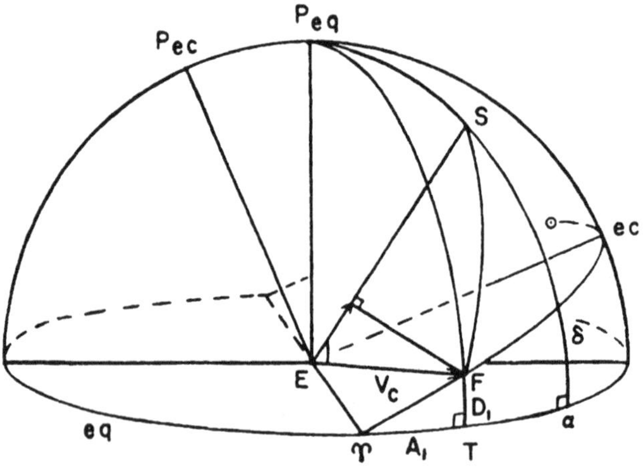

Figure 56. Spherical triangles $P_{eq}SF$ and FT♈.

The underlined terms are newly introduced. We have now:

$$-V_y = 29.76\,(a \sin \odot - b \cos \odot) \qquad (137)$$

where the constants are:

$$\left.\begin{aligned} a &= \cos \alpha \cos \delta \\ b &= \sin \epsilon \sin \delta + \cos \epsilon \sin \alpha \cos \delta \\ &= 0.3979 \sin \delta + 0.9174 \sin \alpha \cos \delta \end{aligned}\right\} \qquad (138)$$

There is another way to derive the same result, namely by substituting the coordinate transformation into the formula for λ, β. In Figure 57 we have in $\Delta\, SP_{eq}P_{ec}$:

$$\left.\begin{aligned} \sin \beta &= \cos \epsilon \sin \delta - \sin \epsilon \cos \delta \sin \alpha \\ a = \cos \beta \cos \lambda &= \cos \alpha \cos \delta \\ b = \cos \beta \sin \lambda &= \sin \epsilon \sin \delta + \cos \epsilon \cos \delta \sin \alpha \end{aligned}\right\}$$

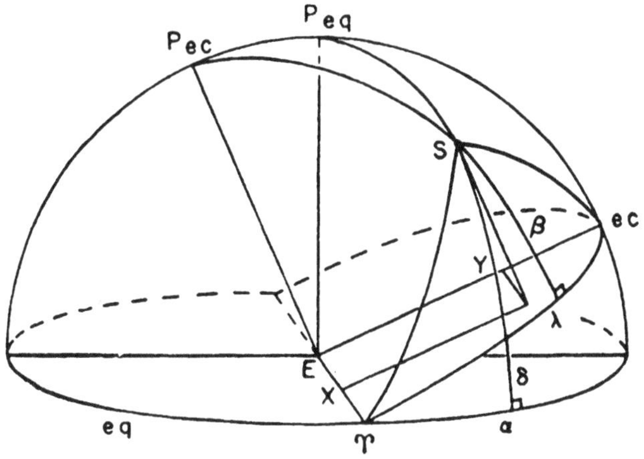

Figure 57. The spherical triangle $P_{eq}P_{ec}S$.

These equations are the direction cosines of the line of sight to the star.

$$\cos \beta \sin (\odot - \lambda) = \cos \beta \sin \odot \cos \lambda - \cos \beta \cos \odot \sin \lambda$$
$$= \cos \alpha \cos \delta \sin \odot - (\sin \epsilon \sin \delta + \cos \epsilon \sin \alpha \cos \delta) \cos \odot$$

For this yearly correction we get again formula (137).

We now consider the actual case of the earth's elliptical orbit. Analogous to our previous result we can predict the result without any computation:

$$- V_y = 29.76\,(a \sin \odot - b \cos \odot) - 0.50\,(a \sin \bar{\omega} - b \cos \bar{\omega}) \tag{139}$$

It is possible to write the formula for the elliptical motion in a simpler notation. S. Herrick Jr. has given the heliocentric velocities of the earth in the plane of the ecliptic for different dates. These are the negative velocities of the sun (Figure 58).

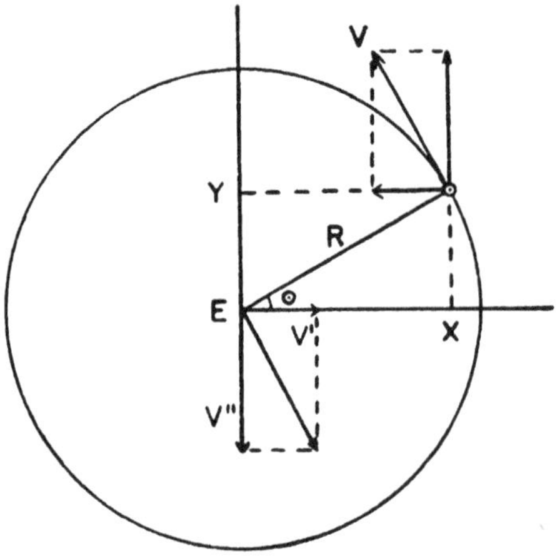

Figure 58. Velocity components in the plane of the ecliptic.

$$\left.\begin{array}{lll} X = R \cos \odot, & \dfrac{dX}{dt} = - V \sin \odot, & V' = + V \sin \odot \\ Y = R \sin \odot, & \dfrac{dY}{dt} = + V \cos \odot, & V'' = - V \cos \odot \end{array}\right\} \tag{140}$$

where X, Y are coordinates and V', V'' are velocities. The expression for the yearly correction now becomes:

$$\begin{aligned} - V_y = {} & (\cos \alpha \cos \delta)\, V \sin \odot \\ & - (\sin \epsilon \sin \delta + \cos \epsilon \sin \alpha \cos \delta)\, V \cos \odot \end{aligned} \tag{141}$$

or introducing the same constants as before:

$$- V_y = aV' + bV'' \qquad (142)$$

In this formula the eccentricity of the earth is thus taken into account. This correction can be easily calculated with help of the computing machine.

We are now able to reduce the measured radial velocity to a heliocentric one which is what is always meant when radial velocity is mentioned in the next paragraphs. The correction to make the time heliocentric is called light time ànd we will consider it in chapter V, paragraph 85. This correction must be applied for spectroscopic double stars with periods smaller than a few days and also for the comparison of the radial velocity curve and the light curve observed during the same night.

47. *Variable radial velocity.* After applying these corrections the radial velocity is the change in distance between the sun and the star per second. Part of this is thus caused by the component of the linear space motion of the sun (20 km/sec) towards the apex.

If the radial velocity of a star is found to be constant during different observing nights, we are sure that it is a single star. If the radial velocity varies with time and only one system of absorption lines is visible, the star may be a single pulsating star or a spectroscopic double star for which one of the components is very faint. One then inspects the spectral type. If the number of lines on comparable plates changes or especially if the spectral type changes with time, the star is single but pulsating (cluster star, Cepheid, red variable). Sometimes these changes are slight and difficult to establish but the pulsating star will usually be indicated by the variation of its light. If the spectral type remains the same with time, the star is probably a spectroscopic double.

On the plates of some stars two spectra appear periodically; thus the absorption lines are sometimes single and sometimes double but in a periodic way. In this case both components of the double have comparable brightness. A spectroscopic double star may show constant light or be an eclipsing star in which case the light curve is usually quite different from that of a pulsating star.

The spectrum variables are peculiar *A* stars which show a periodic change in the depth and shape of the absorption lines, which may be different for different elements. Naturally, in comparing two plates one must realize that the exposure times may have been different. Therefore one makes comparison with a line in the spectrum which keeps a depth constant with time. One can rather easily estimate the ratio of such a line pair or trace the contours of the lines. This change in intensity is often related to variable radial velocity and variable magnetic field, this field causing the widening of the lines.

48. *Equivalent width.* The absorption lines of a star and of our sun originate mainly in the reversing layer of the chromosphere. According to the three Kirchhoff laws it follows that there is a cooler layer surrounding a hotter inner layer called the photosphere. The continuous radiation of the photosphere must pass through the reversing layer. The atoms of the reversing layer absorb those monochromatic wavelengths which they are able to emit themselves; the light in the other wavelengths passes undisturbed. The absorbed radiation is immediately reemitted in all directions. In the direction of the spectroscope one therefore receives only a little radiation in this particular wavelength. One observes an absorption line which is not completely black in the center. The rigorous explanation is somewhat more complicated.

A tracing is made in a machine where a constant light beam

falls on the negative spectral plate which moves with a constant speed perpendicular to the beam. The transmitted beam can fall on a photocell and the resulting photo-electric current can then be recorded on a suitable recording instrument. The density of the spectrum plate is roughly proportional to the monochromatic magnitude over a limited range. A number of density marks of known intensity ratios are also impressed on the plate. A reduction curve can then be constructed from these measurements, and the observed data can be converted into intensities.

The record or tracing of an absorption line could appear as in Figure 59. The shape or contour of a stellar absorption line

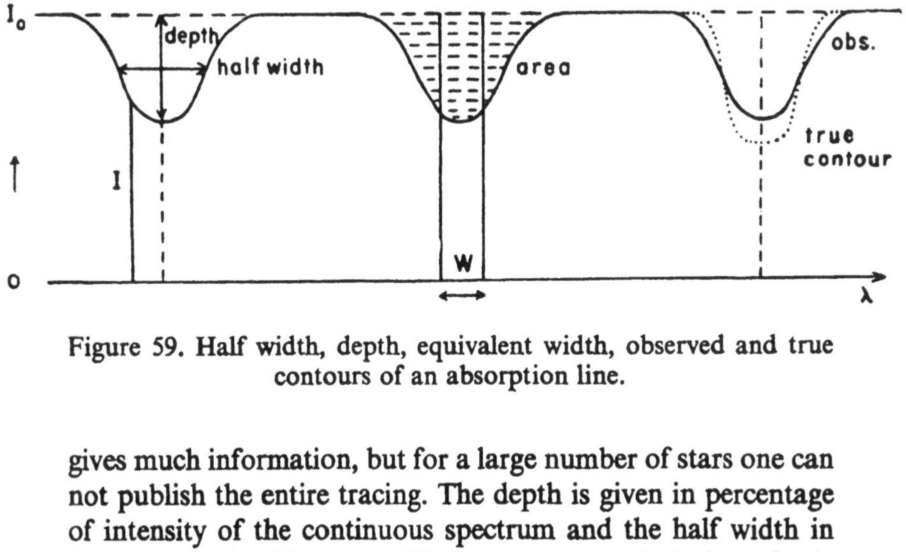

Figure 59. Half width, depth, equivalent width, observed and true contours of an absorption line.

gives much information, but for a large number of stars one can not publish the entire tracing. The depth is given in percentage of intensity of the continuous spectrum and the half width in Angstrom units. However, this is not too practical. A perfectly sharp line would already be smeared out by the finite width of the slit of the spectrograph, and the diffraction pattern of the optics of the spectrograph. Further effects are the diffraction pattern of the collimating lenses used in the microphotometers, as well as the finite slit image in these photometers. Sharp lines are available in the laboratory, and with help of these the so-called apparatus constant can be determined. The true contour of the line can then be computed from the observed one. However, temperature and flexure changes during the observing time also have systematic effects which produce difficulties.

The equivalent width (W) does not have those difficulties. This is defined as the area of the absorption line:

$$W = \int \frac{I_0 - I_\lambda}{I_0} d\lambda = \int \left(1 - \frac{I_\lambda}{I_0}\right) d\lambda \qquad (143)$$

Here I_0 is the intensity in the continuum and I_λ is the intensity at wavelength λ. Another definition of W is the base line in Angstroms of the rectangle which has the same area as the line contour. The dispersion affects the W slightly so that one has to publish the dispersion used.

49. *Curve of growth.* The equivalent width of an absorption line of a star increases with the number of effective atoms in the line of sight. Let N be the number of atoms. The oscillator strength (f) gives the percentage of effective atoms. For the very faint lines the equivalent width increases linearly with the number of effective atoms thus $W \propto Nf$. But if more and more atoms become involved saturation sets in and the equivalent width increases less than the linear relation. If still more atoms are involved, wings appear and the equivalent width again increases by another process. For very strong lines we have $W \propto \sqrt{Nf}$. We are considering here the same absorption line but with different equivalent widths. We can also consider more lines of the same element; in Figure 60 we then use as ordinate W/λ.

The oscillator strength can be computed for all the lines of the simple elements like hydrogen and helium. It can be measured in the laboratory for the more complicated ones. The doublets, triplets and multiplets are especially useful in constructing the curves of growth for the element. It is easy then to compute the relative values of f. For example there are twice as many atoms involved in the formation of the D_2 line of sodium as in the D_1 line. Observation of the equivalent widths shows us directly the effects on the absorption line. We thus find $\log W_1$ and $\log W_2$, and $\log f_2 - \log f_1 = \log 2$ for this

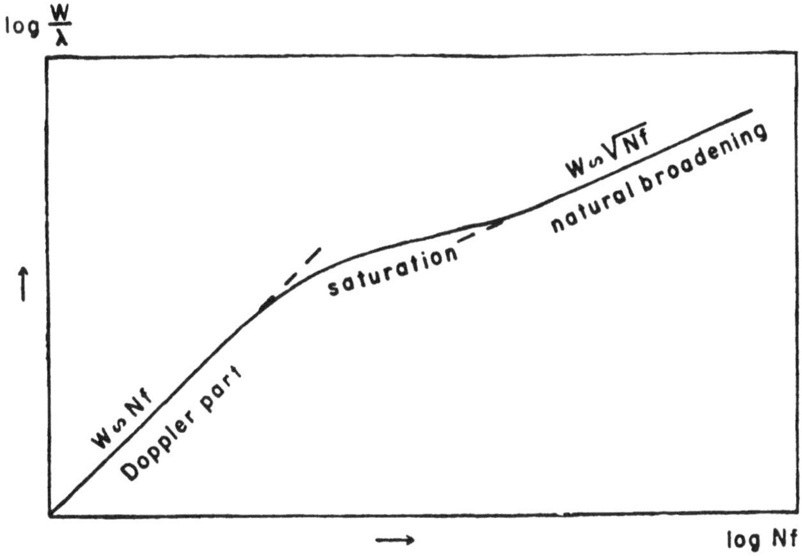

Figure 60. The curve of growth.

segment of the curve of growth for the D lines considered. The slope is known but not yet the actual position, and one can still shift this segment in the x direction. Starting with very faint lines in the left part of the figure we can build up this curve with overlapping segments of other multiplets, For very faint lines the zero point in the Nf is known from experiments in the laboratory (of the order of 10^{12}). During the construction of the curve of growth we assume that W is a single valued function of N, which is sufficiently near the truth. It is clear that the relation found tells us much about the abundances of elements in star atmospheres. One can therefore compare these abundances with those on earth. It is found that hydrogen is by far the most abundant element in the universe.

50. *Spectral classification in two dimensions.* Historically the spectra were classified according to a sequence O, B, A, F, G, K, M and each class was subdivided into ten parts. This is more or less dependent on the number of lines available for the same exposure time and seeing conditions. If one takes the sequence in reversed order, giving each interval M–K, K–G etc. equal scale, we find a rough proportionality with $\log T_e$. For late type stars the effective temperature is low and many more complicated lines appear. For the early type stars the temperature is very high and only hydrogen and helium occur together with a few ionized elements; there are thus only a few lines (Figure 61). Temperature affects most of the absorption lines but in a different way. In determining this classification one considers a number of line pairs. For example the sun is of type $G2$. The calibration between spectral type and absolute temperature can be done with the temperature scale of G. P. Kuiper. The zero point has been determined by the temperature of the sun which can be computed from the heat received on the earth. Per cm^2 we receive 1.94 cal per minute outside the atmosphere; this value is called the solar constant. Per m^2 we receive 323 cal per second and this radiation went through an area of about $(5\ mm)^2$ on the solar surface if one takes the center of the sun as the energy source. This known amount of energy corresponds to a well determined temperature as we will see in chapter V, paragraph 73.

Some of the lines in the spectra are affected by the pressure in the star's atmosphere which in turn depends on the stellar

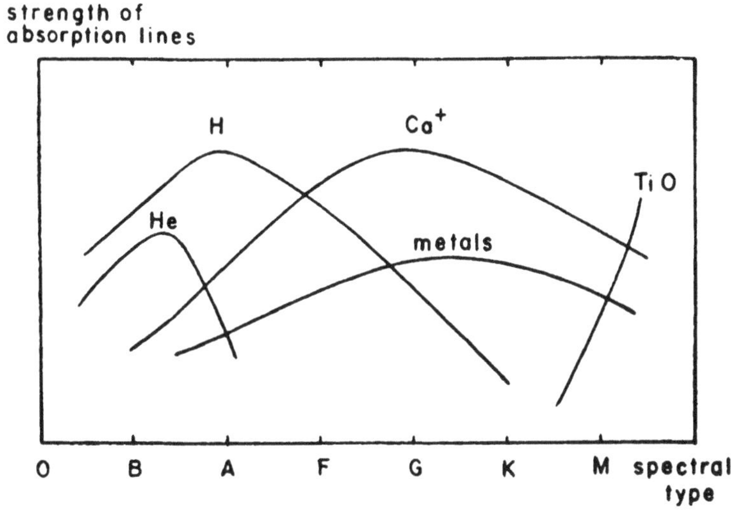

Figure 61. The approximate change in the strength of the absorption lines of certain elements with the spectral type.

size and mass and thus also the absolute magnitude. The hydrogen lines are sharper in giants than in dwarfs but this line sharpness is a qualitative rather than a quantitative criterion. Some lines are stronger in the giants; others in the dwarfs. The strength of a line sensitive to absolute magnitude is estimated with respect to a neighboring nonsensitive line and we thus again consider a line pair or several line pairs. Suitable line pairs are described in the atlas of W. W. Morgan, P. C. Keenan and E. Kellman for the different spectral classes. The observer should first observe the standard stars given in this catalogue because the most suitable criteria may differ somewhat with the various dispersions used. For hot *B* stars R. M. Petrie found the equivalent width of $H\gamma$ to be a good criterion (Figure 62).

For cool stars the cyanogen band provides a useful criterion. Nowadays we can separate five different groups of luminosity (I, II, III, IV, V) over part of the spectral range. It is found that the normal color of giants is redder than the color of dwarfs for the same spectral type, but the observed color is also affected by the interstellar reddening. The strength of the interstellar lines is usually more for giants than for dwarfs because giants may be more distant. In this notation the spectral type of the sun is *G*2V so that the classification is two dimensional.

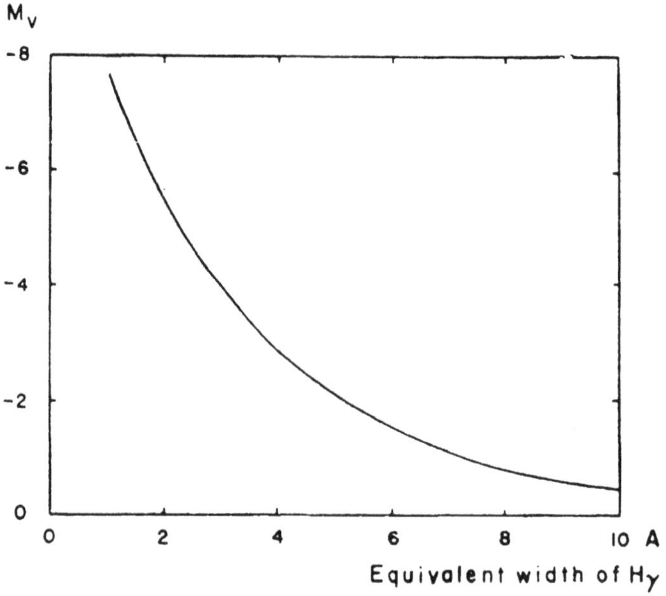

Figure 62. The relation between the equivalent width of $H\gamma$ and the absolute magnitude.

The calibration between luminosity class and absolute magnitude is done with help of stars for which the trigonometric parallax and hence the absolute magnitude is known. For the supergiants and bright giants the distance is so large that we must use cluster parallaxes. If this calibration is done correctly, we can take the absolute magnitude as ordinate and shift every subgroup up to its correct height in the Hertzsprung-Russell diagram (Figure 63). The probable error is about ± 0.2 of a spectral class and ± 0.4 in absolute magnitude.

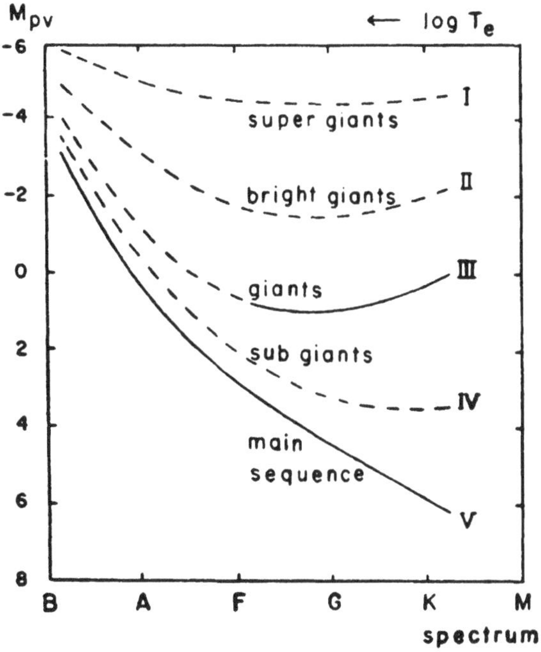

Figure 63. The visual absolute magnitude of the five luminosity classes. Not many stars are found on the dashed lines.

Recently there has been an important increase in accuracy. W. R. Hossack measured the ratio of the line depths on the spectrum plate by electronic means. After calibration with the Morgan system both the spectral class and the luminosity class are given to one more significant figure than before.

A photo-electric method of spectral classification in two dimensions has been reported by B. Strömgren. With help of interference filters of very narrow width (100 A) several points in the spectrum are singled out. For *B*, *A*, and *F* type stars the measurements are made at 3600 A near the Balmer discontinuity with comparisons in the continuum at 4030 A and 4500 A. This comparison gives the Balmer jump which depends on the number of ionized atoms and thus on the temperature. Measurements are made of the $H\beta$ absorption line 4861 A, with comparisons in the continuum on both sides. In a supergiant the hydrogen lines are sharp; in a main sequence star the lines are broad. The effect therefore depends on the pressure in the atmosphere and thus on luminosity class. After calibration with the Morgan classification the accuracy for one observation turns out to be $\pm$ 0.02 of a spectral class and $\pm$ 0.2 in absolute magnitude. For the later type stars other criteria are used. For the determination of the spectral class one then measures the strength of the *K* line (3910 A/4030 A) and the discontinuity at the *G* band (4360 A/4240 A); for the luminosity class one measures the strength of the cyanogen band (4240 A/4170 A). This is a considerable gain in accuracy and means that we will have to introduce subdivisions of one-hundredth of a spectral class. The results are not affected by interstellar absorption.

51. *Spectroscopic parallax.* The pioneer work was done by W. S. Adams and A. Kohlschütter. We will consider here the important problem of calibrating the spectroscopic parallaxes with help of the trigonometric parallaxes. For a given apparent magnitude the absolute magnitude is proportional to the logarithm of the distance or parallax. Therefore for this given apparent magnitude, equal intervals in absolute magnitude do not agree with equal intervals in distance or parallax. In the calibration we must allow for this.

Let us consider as an example the spectral plates of the stars in one of our subgroups. On each plate we can write the information we know about the particular star, namely apparent magnitude m, the trigonometric parallax π_t with its probable error ϵ_t. For this star we can now compute the absolute magnitude M_1 which is provisional because there is a probable error in the trigonometric parallax. This indicates thus an interval around π_t. Going back from M_1 to the parallax should give us our previous starting point. However, this does not give the same result if we consider the spectral plates of the subgroup. We have to compare thus the $\bar{\pi}_t$ with the $\overline{M}_1$ and according to the definition now compute $\bar{\pi}_1$ (see Figure 64).

Let M be the true absolute magnitude, S the systematic error and T the probable error in the provisional magnitude M_1.

$$\begin{aligned} M = M_1 + S + T &= m + 5 + 5 \log (\pi_t + \epsilon_t) \\ M_1 \qquad\qquad &= m + 5 + 5 \log \pi_1 \end{aligned} \tag{144}$$

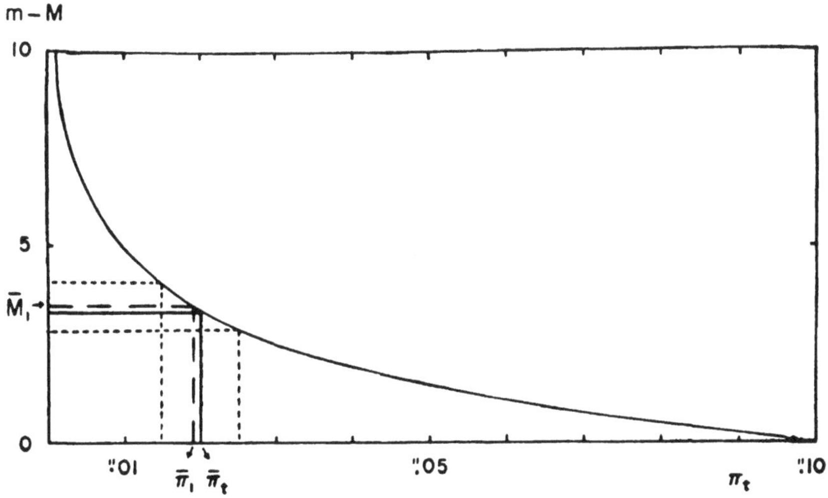

Figure 64. Relation between parallax and distance modulus.

The last line is simply the definition of M_1 (for the derivation see chapter V, paragraph 71); the π_1 is the provisional parallax. If we also divide by 5 subtraction gives:

$$0.2\,S + 0.2\,T = \log \frac{\pi_t + \epsilon_t}{\pi_1}$$

We then find:

$$10^{0.2S}\,10^{0.2T} = \sigma\tau = \frac{\pi_t + \epsilon_t}{\pi_1}$$

We define thus:

$$10^{0.2T} = \tau = e^{0.2T/\text{mod}} = e^E = 1 + E + \tfrac{1}{2}E^2 + \ldots (E \text{ small})$$

$$E = \frac{0.2T}{\text{mod}} \qquad \text{Further let:} \qquad 1 + \tfrac{1}{2}E^2 = A \geqslant 1$$

$$\pi_1\sigma = \frac{\pi_t + \epsilon_t}{\tau} = (\pi_t + \epsilon_t)(1 - E + \tfrac{1}{2}E^2) = (\pi_t + \epsilon_t)(A - E)$$

Working this out we can take: $\epsilon_T A \approx \epsilon_T$ and $\epsilon_T E = 0$

$$\pi_1\sigma - \pi_t A = \epsilon_t - \pi_t E \tag{145}$$

This last equation is valid for the data of any star in the subgroup. For all stars in the subgroup we have now:

$$\bar{\pi}_1\sigma - \bar{\pi}_t A = 0, \qquad \frac{\sigma}{A} = \frac{\bar{\pi}_t}{\bar{\pi}_1} \tag{146}$$

If we give weights to the individual stars we find:

$$\frac{\sigma}{A} = \frac{\Sigma p\,\pi_t}{\Sigma p\,\pi_1} \tag{147}$$

This ratio can now be computed for the subgroup. The calibration between the absolute magnitude and the parallax must hold for the subgroup in such a way that the mean spectroscopic parallax agrees with the mean trigonometric parallax for all the stars in the subgroup.

$$\bar{\pi}_{sp} = \bar{\pi}_t = \bar{\pi}_1 \frac{\sigma}{A} \tag{148}$$

The best value for any star in the subgroup is thus:

$$\pi_{sp} = \pi_1 \frac{\sigma}{A} \tag{149}$$

The correction factor σ/A has already been computed. In absolute magnitude it corresponds to the term $5 \log (\sigma/A)$. After applying this correction we have calibrated our system.

Observing the spectral plate of another star we determine the spectral classification and can place its data in the correct subgroup. We therefore immediately know its absolute magnitude. From the apparent magnitude m_0 corrected for interstellar absorption we find now π_{sp}. There is a probable error in the absolute magnitude, thus also in the $\log \pi_{sp}$ or in the $\log r$. This means that there is a percentage probable error in the π_{sp} and in the r. For example a probable error of $\pm$ 0.3 in the absolute magnitude gives a probable error of $\pm$ 0.06 in the $\log \pi_{sp}$ or $\log r$, which corresponds to a 15% probable error in π_{sp} or in r.

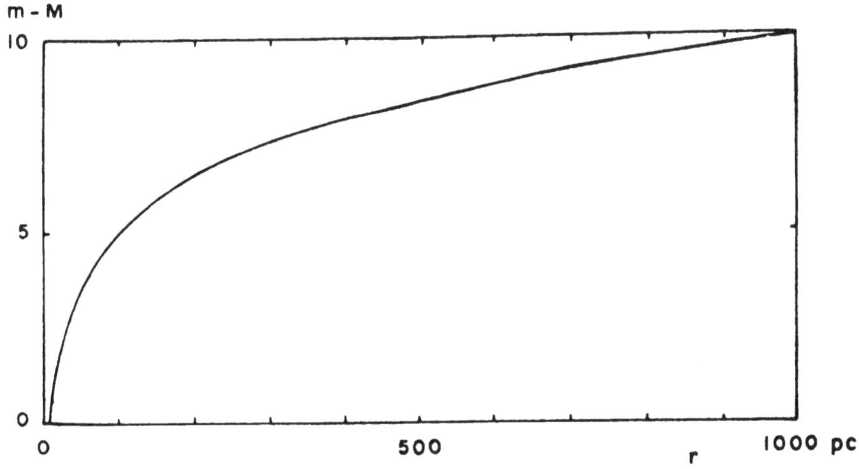

Figure 65. Relation between distance and distance modulus.

This is a great advantage for very distant stars for which the trigonometric parallax with the constant probable error no longer gives reliable results. The spectroscopic parallax is useful for distances between 40 parsecs and 2000 parsecs. The distance modulus is defined as:

$$m_0 - M = -5 \log \pi_y - 5 = 5 \log r - 5 \qquad (150)$$

where m_0 is the apparent magnitude corrected for reddening, M is the absolute magnitude for distance 10 parsecs, π_y is the yearly parallax in seconds of arc, r is the distance in parsecs (see Figure 65). For example, a distance modulus of 5 corresponds to a distance of 100 parsecs.

52. *Peculiar spectra.* There are a number of peculiar spectra, which do not fit into this two dimensional diagram. For example the metallic line stars give different spectral classifications if one classifies them according to different line pairs.

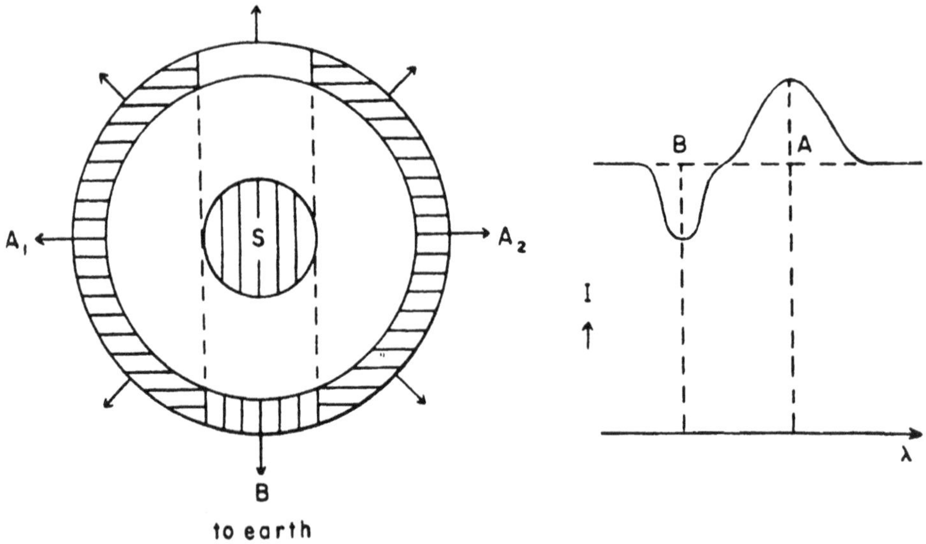

Figure 66. Expanding shell and contour of a spectral line of a nova.

A so-called shell spectrum belongs to a star with an outer atmospheric envelope. One type of shell star is the so-called P Cygni type, which is surrounded by an expanding atmosphere. Novae also possess this type of spectrum during early post maximum stage. Ring parts A_1 and A_2 in Figure 66 give the emission line which is widened but undisplaced; part B gives the absorption line which is displaced to the violet side of the bright line. The outer shell here conceals the structure of the star inside.

Many *B* stars have emission lines. They are called *Be* stars, and have a rotating shell. The star itself gives the broad absorption lines S in Figure 67. The two ring parts give the emission line A; part B produces a narrow absorption line in this emission line. The total contour is therefore rather complicated. With axes at various angles to the line of sight, we get wide or narrow emission. Strongest absorption occurs in those whose equators lie exactly in the line of sight.

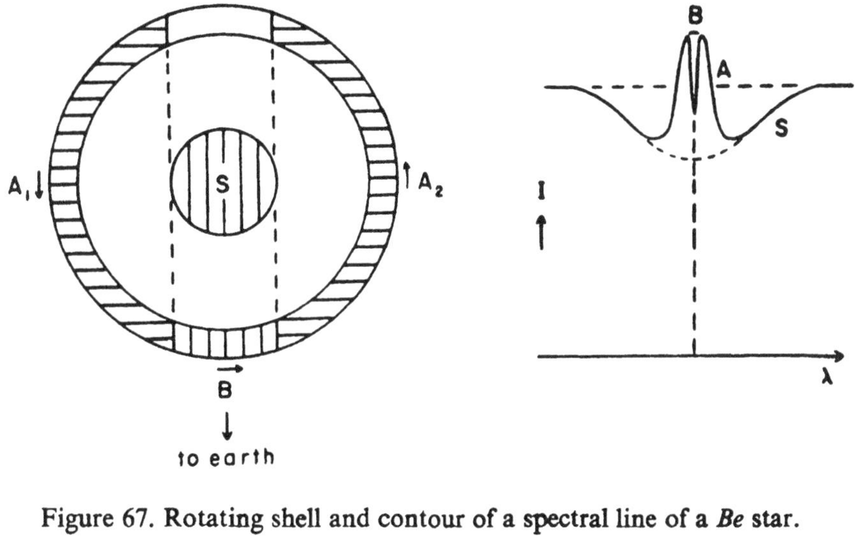

Figure 67. Rotating shell and contour of a spectral line of a *Be* star.

53. *Interstellar lines.* These are very sharp lines which do not participate in the movement of the star, as can be shown in case of a spectroscopic double star. In fact they were first found in this way by J. Hartmann. Such interstellar lines may be shifted from the rest position and have a constant radial velocity. They become deeper for stars at larger distances. The explanation is that they are caused by interstellar gases between the star and the observer (Figure 68).

Figure 68. Interstellar and stellar absorption lines.

The *O* and *B* stars are very favorable objects to use in a study of these lines since in the early type spectra there are relatively few stellar lines. Further the absolute luminosity of these stars is so large that we can observe them at large distances. Under the assumption that the distribution of the interstellar gas is at random, the equivalent width is a measurement of the distance of the star. It is now known that there is some structure in these gases so that this method is valid only in a small area. Sometimes the lines are double or multiple meaning that there are two or more clouds in the line of sight. Also some molecular interstellar bands have been observed. Most observations are made of the *K* and *H* doublet of Ca^+, the D_1 and D_2 doublet of *Na*, and the λ 4430 band. The strength of this last band is proportional to the reddening.

54. *Parallax from differential galactic rotation.* The center of our galactic system is in the constellation of Sagittarius. This is found from the center of the system of globular clusters which

for dynamical reasons must coincide with the center of the galaxy. Red and infrared plates show many red giants in this nucleus, and radio observations show a strong maximum in its direction of galactic longitude $l_0 = 325°$.

This entire system is rotating. The sun for example is moving in the direction of the constellation of Cygnus 90° from the center direction with a speed of 210 km/sec. This is found from radial velocities of the globular clusters using the system of clusters as a reference and from radio observations. For simplicity we take a circular orbit for the sun around the center of the Milky Way. Another star may move at a somewhat different speed. The difference shows up as the proper motion and radial velocity, and thus the space motion of a star relative to the sun.

The sun must be close to the plane of the galaxy because we see the Milky Way along a great circle in the sky. It has to be near the edge because in the anticenter direction our line of sight soon passes out of the system and we do not see too many stars. Further it is situated in an outer faint spiral arm of the system.

For the outer regions rotation is similar to that of a planetary system where outer objects move slower than do inner ones. The nucleus can then be taken as the mass center. We first correct our data for the linear solar motion of 20 kilometers per second with respect to the nearby stars.

If the circular speed of the sun is V, then the speed of another star with polar coordinates r, $\theta = l - l_0$ is (Figure 69):

$$V + dV = V + \frac{\partial V}{\partial R} dR = V - r \cos \theta \frac{\partial V}{\partial R} \tag{151}$$

Here R is the distance from the center to the sun. Because r is small compared with R, in $\Delta \odot SC$ we have simply:

$$\cos \varphi = 1, \quad \frac{\sin \varphi}{\sin \theta} = \frac{r}{R}$$

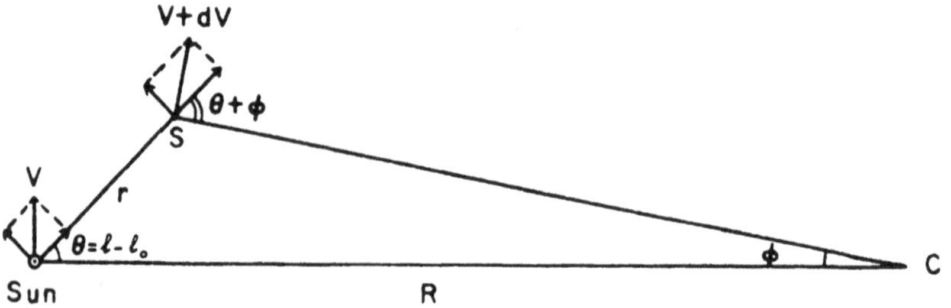

Figure 69. Velocity components in the plane of the Milky Way.

The observed radial velocity is the difference of the two motions projected in the line of sight. We have (Figure 70):

$$V_r = \left(V - r\cos\theta\frac{\partial V}{\partial R}\right)\sin(\theta + \varphi) - V\sin\theta$$

$$= \left(V - r\cos\theta\frac{\partial V}{\partial R}\right)\left(\sin\theta + \frac{r}{R}\cos\theta\sin\theta\right) - V\sin\theta$$

$$= \frac{Vr}{R}\cos\theta\sin\theta - r\cos\theta\sin\theta\frac{\partial V}{\partial R}$$

$$= r\left(\frac{V}{R} - \frac{\partial V}{\partial R}\right)\cos\theta\sin\theta = \frac{r}{2}\left(\frac{V}{R} - \frac{\delta V}{\delta R}\right)\sin 2\theta$$

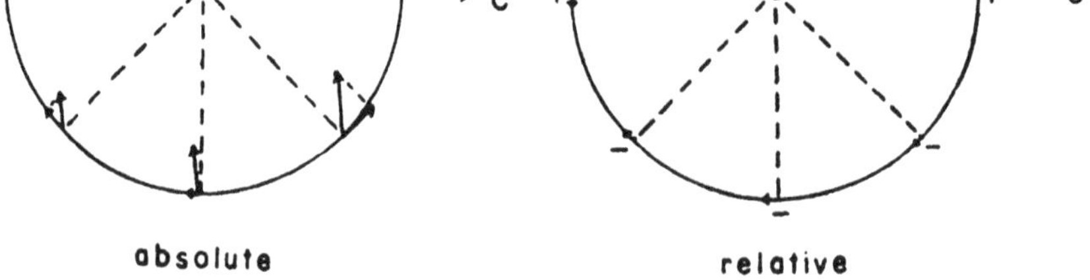

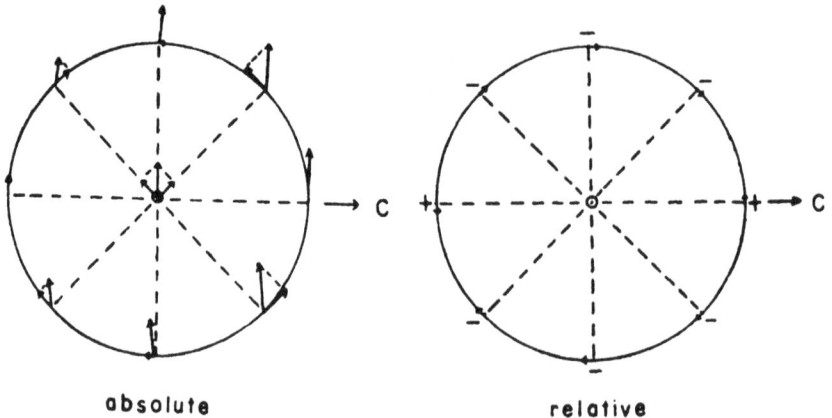

Figure 70. Absolute and relative motions in the plane of the Milky Way. The relative radial velocities are given as observed.

We can write this result as follows (Figure 71):

$$V_r = rA \sin 2(l - l_0), \qquad A = \tfrac{1}{2}\left(\frac{V}{R} - \frac{\partial V}{\partial R}\right) \tag{152}$$

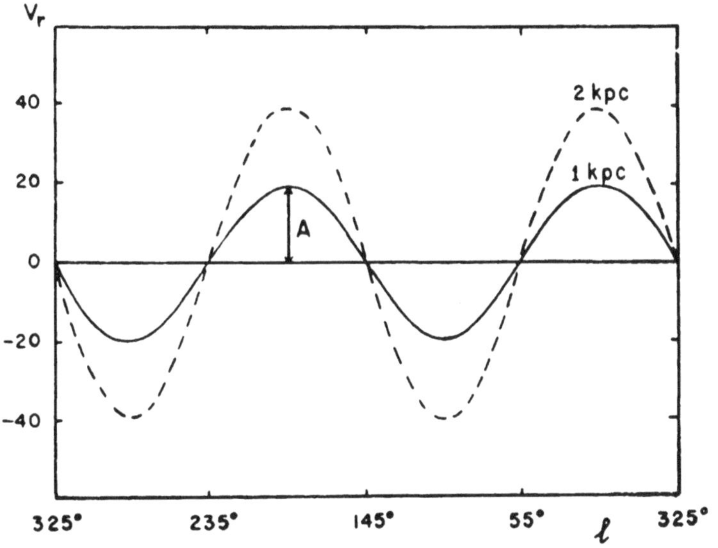

Figure 71. Differential galactic rotation effect in the radial velocities given for two distances to the sun.

This is in the plane of the Milky Way. For galactic latitude b we see that in Figure 72:

$$V_r = rA \sin 2(l - l_0) \cos b, \text{ where } r = r' \cos b$$

If now we call r' the distance to a star in latitude b we find:

$$V_r = r'A \sin 2\,(l - l_0) \cos^2 b \tag{153}$$

This is a double sine wave. There are four zero points or nulls, namely the center, anticenter, and the two intermediate points in the plane of the Milky Way where the differential effect vanishes. To a first approximation the amplitude is proportional to the distance of the stars. From stars of known distance one can find the curve and determine A and l_0. Now we take a star of unknown distance which is situated around the extremes, and thus is far from the zero points. The radial velocity, freed from the linear solar motion, is then caused by differential galactic rotation effect and the peculiar motion of the star in the line of sight. For a group of stars this last effect is assumed to be zero

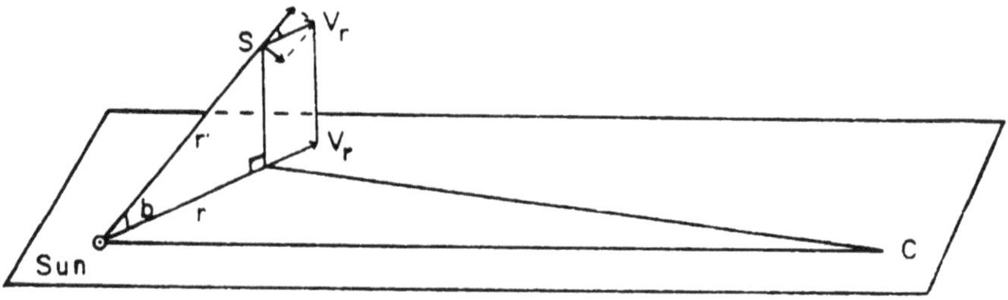

Figure 72. Velocity components for a star with galactic latitude b.

in the mean and we find the distance of the group. This method is more accurate if the stars are far away, up to 2000 parsecs, because then the galactic rotation is large. For very distant stars additional terms have to be added to the formula; these distort the curve a little.

Interstellar material also exhibits differential galactic rotation. For the nearby stars it is found that the effective distance of this material is about half the stellar distance. Both a uniform distribution and a cloud distribution at random will produce such a result. If very distant stars in the anticenter region are observed two very pronounced interstellar lines are found. They correspond to the two spiral arms in that direction. With formula (153) we can now find the distance of these two spiral arms and locate them. This is done in a similar way with radio observations of the hydrogen line of 21 cm wave length. Again from the Doppler shift of the radial velocities caused by the differential galactic rotation effect we find the distance of a spiral arm in that direction.

In the tangential plane we find in kilometers per second (Figure 73):

$$\begin{aligned}
V_t &= \left(V - r\cos\theta\frac{\partial V}{\partial R}\right)\cos(\theta+\varphi) - V\cos\theta \\
&= \left(V - r\cos\theta\frac{\partial V}{\partial R}\right)\left(\cos\theta - \frac{r}{R}\sin^2\theta\right) - V\cos\theta \\
&= -\frac{Vr}{R}\sin^2\theta - r\cos^2\theta\frac{\partial V}{\partial R} \\
&= -\frac{Vr}{R}(1-\cos^2\theta) - r\cos^2\theta\frac{\partial V}{\partial R} \\
&= -\frac{Vr}{R} + r\left(\frac{V}{R} - \frac{\partial V}{\partial R}\right)\cos^2\theta \\
&= -\frac{Vr}{R} + \frac{r}{2}\left(\frac{V}{R} - \frac{\partial V}{\partial R}\right) + \frac{r}{2}\left(\frac{V}{R} - \frac{\partial V}{\partial R}\right)\cos 2\theta \\
&= -\frac{r}{2}\left(\frac{V}{R} + \frac{\partial V}{\partial R}\right) + \frac{r}{2}\left(\frac{V}{R} - \frac{\partial V}{\partial R}\right)\cos 2\theta
\end{aligned}$$

This can be written as:

$$V_t = rB + rA\cos 2(l - l_0), \qquad B = -\tfrac{1}{2}\left(\frac{V}{R} + \frac{\partial V}{\partial R}\right) \tag{154}$$

This is the expression in the Milky Way plane. For galactic latitude b we get:

$$V_t = r\{A\cos 2(l - l_0) + B\}\cos b$$

One astronomical unit per year corresponds to 4.74 km/sec. The proper motion in seconds of arc per year is $\mu_l = V_t/4.74r$. In the plane of the Milky Way we have:

$$\mu_l = \frac{A}{4.74}\cos 2(l - l_0) + \frac{B}{4.74}$$

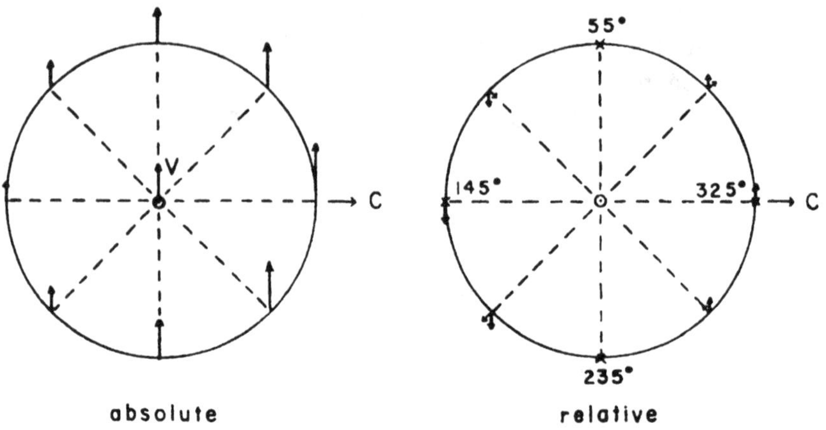

Figure 73. Absolute and relative motions in the plane of the Milky Way. The relative proper motions are given as observed.

We again have a double cosine wave but shifted in the ordinate (Figure 74). B depends on V, R, $\partial V/\partial R$. For latitude b we find:

$$\mu_l \cos b = \left\{\frac{A}{4.74} \cos 2(l - l_0) + \frac{B}{4.74}\right\} \cos b \qquad (155)$$

The $\mu_l \cos b$ is independent of the distances of the stars.

The constants A and B are called the Oort constants, because J. H. Oort determined them first from observations. Note that:

$$A - B = \frac{V}{R}, \qquad -(A + B) = \frac{\partial V}{\partial R} \qquad (156)$$

The following values are found for the constants:

$$\left.\begin{aligned} A &= +0.0195 \text{ km/sec.pc}, & \frac{A}{4.74} &= +0\farcs0041 \\ B &= -0.0069 \text{ km/sec.pc}, & \frac{B}{4.74} &= -0\farcs0015 \end{aligned}\right\} \qquad (157)$$

From this we derive:

$$\left.\begin{aligned} A - B &= +0.0264 \text{ km/sec.pc} = \frac{V}{R} \\ -(A + B) &= -0.0126 \text{ km/sec.pc} = \frac{\partial V}{\partial R} \end{aligned}\right\}$$

Thus ΔV is negative for positive ΔR. Going outwards the true velocity decreases indicating that the sun is near the edge of the system because only there can this be possible.

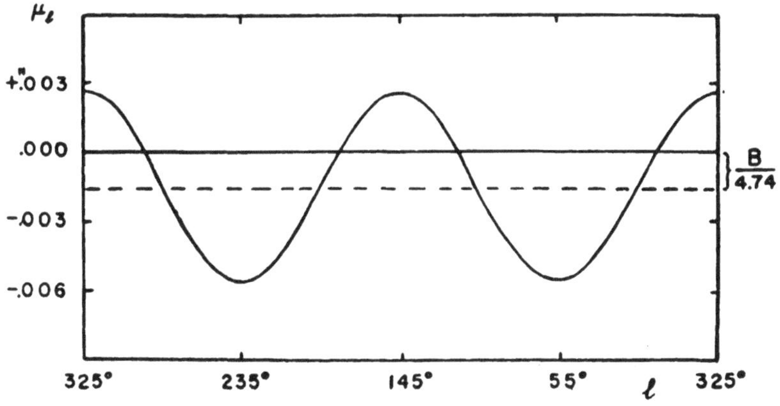

Figure 74. Differential galactic rotation effect in the proper motions.

The average distance of the nucleus of the Milky Way is found to be 8000 parsecs or 8 kiloparsecs from an investigation of cluster variables in a region of the great star cloud in Sagittarius. Substituting this we find $V = 210$ km/sec. For a star 1000 parsecs more towards the anticenter the velocity decreases by 12.6 km/sec.

REFERENCES

SPECTROGRAPH

W. W. Campbell: *Stellar motions.* Yale Un. Press, New Haven, 1913.

STANDARD WAVELENGTHS

Trans. I.A.U., **1**, 39, 1922; **2**, 40, 1925; **3**, 86, 1928; **4**, 58, 1932; **5**, 81, 1935; **6**, 79, 1938; **7**, 146, 1950; **8**, 188, 1952; **9**, 201, 1955.

M.I.T. wavelength tables. John Wiley & Sons, 1939.

REDUCTION

J. Hartmann: *Ap.J.*, **8**, 218, 1898; **24**, 285, 1906.

S. Herrick Jr: *L.O.B.*, **17**, 85, 1935.

RADIAL VELOCITIES

J. H. Moore: *The radial velocities of stars*. Chapter V of R. G. Aitken's: *The binary stars*. McGraw-Hill Book Co., New York and London, 1935.

R. E. Wilson: *General catalogue of stellar radial velocities*. Carnegie Inst. of Washington Publ. No. 601, 1953.

LINE PROFILES

R. O. Redman: *M.N.*, **95**, 290 and 742, 1935.

CURVE OF GROWTH

L. Goldberg and L. H. Aller: *Atoms, stars and nebulae*. The Blakiston Co., Philadelphia, 1943.

Ch. E. Moore: *A multiplet table of astrophysical interest. Princeton Obs. Contr.* No. 20, 1945.

L. H. Aller: *Astrophysics; The atmospheres of the sun and stars*. Ronald Press Co., New York, p. 288, 1953.

ABSOLUTE MAGNITUDE EFFECT

W. S. Adams and A. Kohlschütter: *Ap. J.*, **40**, 385, 1914; *Mt. W. Contr.* No. 89.

B. Lindblad: *Ap. J.*, **104**, 325, 1946.

R. M. Petrie: *R.A.S.C. Journal*, **50**, 49, 1956; *Contr. Dominion Astroph. Obs. Victoria*, No. 34, 1956.

SPECTRAL CLASSIFICATION

W. W. Morgan, P. C. Keenan and E. Kellman: *An atlas of stellar spectra*. Un. of Chicago Press, Chicago, 1943.

W. A. Hiltner and R. Williams: *Photometric atlas of stellar spectra.* Un. of Michigan, 1946.
G. P. Kuiper: *Ap. J.*, **88**, 429, 1938.
W. R. Hossack: *Ap. J.*, **119**, 613, 1954.
B. Strömgren: *A.J.*, **59**, 193, 1954.
B. Strömgren and K. Gyldenkerne: *Ap. J.*, **121**, 43, 1955.

SPECTROSCOPIC PARALLAX

W. S. Adams, A. H. Joy and collaborators: *Ap.J.*, **46**, 313, 1917; **53**, 13, 1921; **64**, 225, 1926; **81**, 187, 1935.
J. Ramsey: *Ap.J.*, **111**, 434, 1950.
D. Duke: *Ap.J.*, **113**, 100, 1951.

INTERSTELLAR LINES

L. Binnendijk: *Ap.J.*, **115**, 428, 1952.
L. Spitzer, Jr.: *Ap.J.*, **120**, 1, 1954.
J. Dufay: *Nébuleuses galactiques et matière interstellaire*, Albin Michel, Paris, 1954; *Galactic nebulae and interstellar matter.* Philosophical Library, New York, 1957.

SPECTRUM VARIABLES

A. J. Deutsch: *Ap.J.*, **105**, 283, 1947.

GALACTIC ROTATION

J. H. Oort: *B.A.N.*, **3**, 275, 1927; **4**, 79, 91, 269, 1927–28.

IV

Spectroscopic Double Stars

A SPECTROSCOPIC DOUBLE STAR IS SEEN SINGLY BY THE naked eye or in the telescope, but the duplicity is discovered by the periodic shift of the spectral absorption lines. Except in certain peculiar systems the spectral type does not change during the period. To get a measurable displacement of the lines the speed in the line of sight has to be at least a few km/sec. The orbit must therefore be rather small since the double generally does not appear as a visual double star. About 1000 systems have been discovered of which some 400 have well determined orbits.

55. *One spectrum visible.* In this case the companion is faint compared with the primary star and we are considering the absolute orbit of the brighter star S_1 around the gravity center. The semi-major axis of this absolute orbit is a_1 (and a_2 for the fainter component) but the subscript is omitted for the methods of orbital determination for the case of one spectrum visible. In this case we can determine only the combination $a \sin i$. The inclination itself thus remains undetermined. Often these stars are also eclipsing variables and then the i can be determined.

Inherent in the problem of the orbital determination is the fact that the position angle of the node Ω will remain unknown. However, if the star is at the node we measure the direction of its speed in the line of sight and can thus tell whether it be the ascending or descending node. Therefore, we can define ω as the angular distance of periastron from the ascending node, measured in the direction of the orbital motion.

In Figure 75 we compare the absolute ellipse of the brighter star with the relative ellipse of the same double star studied

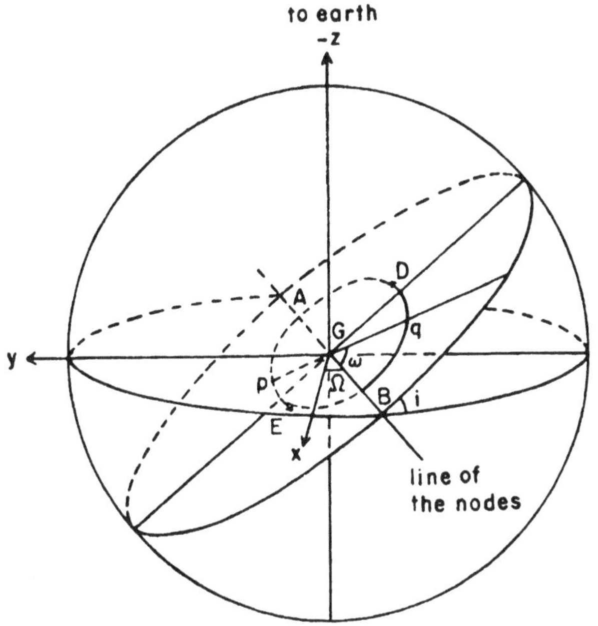

Figure 75. Orbital ellipse in true plane.

previously. The axis $-z$ is directed towards the earth. A comparison of both ellipses now gives:

$$z = -r \sin(v + \omega + 180) \sin i = r \sin(v + \omega) \sin i \qquad (158)$$

To get the speed along the z-axis we have to take the derivative.

$$\frac{dz}{dt} = \sin(v + \omega) \sin i \frac{dr}{dt} + r \cos(v + \omega) \sin i \frac{dv}{dt} \qquad (159)$$

We now have to compute the two derivatives. For dr/dt we start with the equation of an ellipse in polar coordinates:

$$r = \frac{a(1-e^2)}{1+e\cos v}, \qquad \frac{1}{r} = \frac{1+e\cos v}{a(1-e^2)} \tag{160}$$

Differentiating of $1/r$ gives now:

$$\frac{1}{r^2}\frac{dr}{dt} = \frac{e\sin v}{a(1-e^2)}\frac{dv}{dt}$$

$$\frac{dr}{dt} = \frac{r^2 e\sin v}{a(1-e^2)}\frac{dv}{dt} = \frac{h\,e\sin v}{a(1-e^2)} = \frac{2\pi ab}{P}\frac{e\sin v}{a(1-e^2)}$$

$$= nb\frac{e\sin v}{1-e^2} = na\sqrt{(1-e^2)}\frac{e\sin v}{1-e^2} = \frac{nae\sin v}{\sqrt{(1-e^2)}}$$

We used here Kepler's second law which is the starting point for the computation of dv/dt.

$$r^2\frac{dv}{dt} = h = \frac{2\pi ab}{P} = nab$$

$$r\frac{dv}{dt} = \frac{nab}{r} = \frac{nab(1+e\cos v)}{a(1-e^2)} = \frac{nb(1+e\cos v)}{1-e^2}$$

$$= \frac{na\sqrt{(1-e^2)}(1+e\cos v)}{1-e^2} = \frac{na(1+e\cos v)}{\sqrt{(1-e^2)}}$$

We will also use the following formula:

$$\cos\omega = \cos\{(v+\omega) - v\}$$
$$= \cos v\cos(v+\omega) + \sin v\sin(v+\omega)$$

Our expression now becomes:

$$\frac{dz}{dt} = \frac{nae\sin v}{\sqrt{(1-e^2)}}\sin(v+\omega)\sin i$$
$$+ \frac{na(1+e\cos v)}{\sqrt{(1-e^2)}}\cos(v+\omega)\sin i$$
$$= \frac{na\sin i}{\sqrt{(1-e^2)}}\{e\cos\omega + \cos(v+\omega)\}$$

The observed radial velocity consists of a constant part, namely the radial velocity V_0 of the gravity center, and also dz/dt, which is variable with time since the v varies with time. All the other symbols in the formula are orbital constants.

$$V_r = V_0 + K\{e \cos \omega + \cos (v + \omega)\} \tag{161}$$

where half the amplitude is given as:

$$K = \frac{n\,a \sin i}{\sqrt{(1-e^2)}}, \qquad n = \frac{2\pi}{P} \tag{162}$$

For a circular orbit $e = 0$ and thus $K = n\,a \sin i$. This can be seen very easily from Figure 76. The speed in the true orbit is $V_c = 2\pi\, a/P = n\,a$. The orbital plane is imagined to be perpendicular to the plane of the drawing. At D and E the speed is

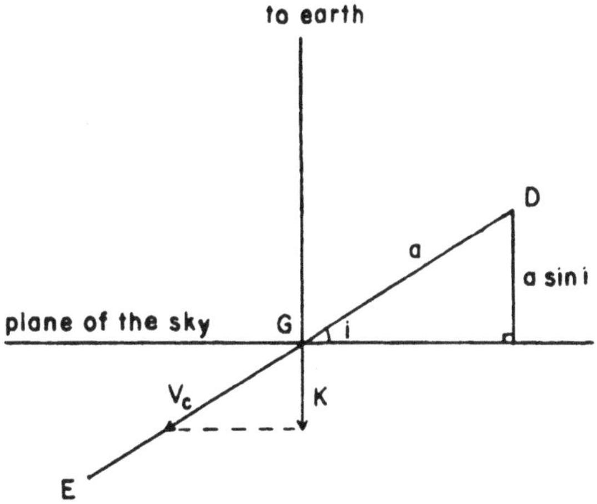

Figure 76. Relation between circular velocity and *K*.

perpendicular to the line of sight, and the radial velocity is there zero. But a quarter of a revolution later this point is projected at G. There, the circular speed gives the maximum or the minimum component in the line of sight and thus in the radial velocity. It is seen that the triangles are similar, from which it follows that:

$$K : na = a \sin i : a, \qquad K = n\, a \sin i$$

Moreover, for a circular orbit the true anomaly v is proportional to time. The radial velocity diagram is here a perfect cosine curve.

56. *Method of R. Lehmann-Filhés.* In the radial velocity diagram the determination of the period P proceeds in the same way as in the case of a light curve. We know that the time interval between two maxima (or minima) equals a whole number of periods so we try to find the smallest interval. After this provisional period has been found, we use the middle of the increasing branch (or decreasing branch) for a sharper determination of P. A least squares solution gives the best result. From this we can find $n = 2\pi/P$. We can remark here that the relation between the observed period and the true period P_0 is given by:

$$\frac{P}{P_0} = 1 + \frac{V_0}{c} \tag{163}$$

We find the V_0 as follows. With respect to the gravity center the bright star will come back to the same orientation after one period; the coordinates z_{t+P} and z_t are the same (Figure 77).

$$z_{t+P} - z_t = \int_t^{t+P} \frac{dz}{dt}\, dt = 0 \tag{164}$$

The ordinate is the radial velocity dz/dt in the velocity diagram, dt being along the time axis. The product is then the area of the

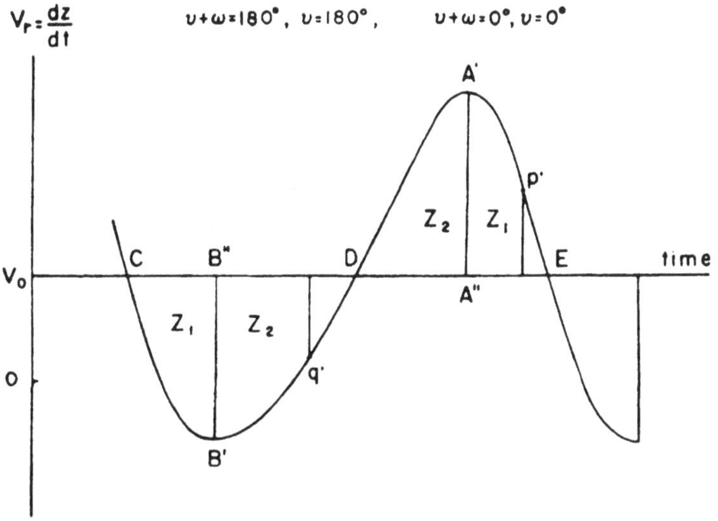

Figure 77. Radial velocity diagram with V_0-axis.

rectangle in this diagram. In space along the line of sight, the difference of coordinates z corresponds to an area in the radial velocity diagram. To make the distinction we will write the areas as Z if we talk about the velocity diagram. At the same time we will compare these areas with the z-coordinates of the absolute orbit of the bright star in space with respect to the center of gravity.

$$\text{area DA'E} = \int_D^E \frac{dz}{dt}\,dt = z_E - z_D$$

$$\text{area CB'D} = \int_C^D \frac{dz}{dt}\,dt = z_D - z_C = -(z_C - z_D)$$

We now choose our V_0 axis in such a way that area DA'E = area CB'D except for the sign. Then $z_E = z_C$ and indeed the z coordinate in space is then the same after one period. With respect to the gravity center this is required for the position of the bright star. We now take this V_0 axis as our coordinate axis so that the equation of the radial velocity curve with respect to the gravity center is:

$$V_r = K\{e \cos \omega + \cos (v + \omega)\} \tag{165}$$

Let A and B be the measured lengths with respect to the V_0 axis, thus numerically positive. Note this especially for the B.

$$\left.\begin{array}{ll} A = K(1 + e \cos \omega) & \text{for } v + \omega = 0^\circ \\ B = K(1 - e \cos \omega) & \text{for } v + \omega = 180^\circ \end{array}\right\} \tag{166}$$

The K is now found as half the amplitude.

$$K = \tfrac{1}{2}(A + B) \tag{167}$$

Further we have:

$$K e \cos \omega = \tfrac{1}{2}(A - B), \qquad e \cos \omega = \frac{A - B}{A + B} \tag{168}$$

We would like to find an expression for $e \sin \omega$, which would enable us to solve for e and ω. During the maximum of the radial velocity curve the bright star passes the node at A while at E the star is as far away as possible. With respect to the center of gravity we now have:

$$Z_1 = \int_A^E \frac{dz}{dt}\, dt = z_E - z_A = z_E \tag{169}$$

$z_A = 0$ because the node is situated on the line of the nodes which in this case goes through the center of gravity.

$$Z_1 = \text{area A'A''E} = \text{area CB'B''} = z_E$$
$$Z_2 = \text{area DA'A''} = \text{area B'B''D} = z_D$$

The best way to find the maximum and the minimum of the radial velocity curve is to connect the middle of chords drawn parallel with the time axis. The ordinate through the maximum now divides the area above the V_0 axis into the areas Z_1 and Z_2, and the ordinate through the minimum does the same for the area below the axis.

At E the star is moving parallel with the line of the nodes; the same is true at D. Thus $dz/dt = 0$, and the formula becomes:

$$0 = e \cos \omega + \cos (v + \omega)$$

We have thus for point E:

$$\left.\begin{aligned} \cos (v_1 + \omega) &= -e \cos \omega = \frac{B - A}{A + B} \\ \sin (v_1 + \omega) &= \sqrt{1 - \left(\frac{B - A}{A + B}\right)^2} = \frac{2\sqrt{AB}}{A + B} \end{aligned}\right\}$$

and for point D:

$$\left.\begin{aligned} \cos (v_2 + \omega) &= -e \cos \omega = \frac{B - A}{A + B} \\ \sin (v_2 + \omega) &= -\frac{2\sqrt{AB}}{A + B} \end{aligned}\right\}$$

The cosine is the same in both cases; the sine is the same except the sign. Further we have:

$$Z_1 = z_E = r_1 \sin (v_1 + \omega) \sin i$$
$$Z_2 = z_D = r_2 \sin (v_2 + \omega) \sin i = -r_2 \sin (v_1 + \omega) \sin i$$

We will also count Z_2 as a positive quantity and have:

$$\frac{Z_1}{Z_2} = \frac{r_1}{r_2} = \frac{1 + e \cos v_2}{1 + e \cos v_1} = \frac{1 + e \cos \{(v_2 + \omega) - \omega\}}{1 + e \cos \{(v_1 + \omega) - \omega\}}$$
$$= \frac{1 + e \cos (v_2 + \omega) \cos \omega + e \sin (v_2 + \omega) \sin \omega}{1 + e \cos (v_1 + \omega) \cos \omega + e \sin (v_1 + \omega) \sin \omega}$$
$$= \frac{1 - \cos^2 (v_1 + \omega) - e \sin (v_1 + \omega) \sin \omega}{1 - \cos^2 (v_1 + \omega) + e \sin (v_1 + \omega) \sin \omega}$$
$$= \frac{\sin^2 (v_1 + \omega) - e \sin (v_1 + \omega) \sin \omega}{\sin^2 (v_1 + \omega) + e \sin (v_1 + \omega) \sin \omega}$$

$$\frac{Z_1}{Z_2} = \frac{\sin (v_1 + \omega) - e \sin \omega}{\sin (v_1 + \omega) + e \sin \omega} \qquad (170)$$

According to a property of ratios we have now:

$$\frac{Z_2 - Z_1}{Z_2 + Z_1} = \frac{2\,e \sin \omega}{2 \sin (v_1 + \omega)} = \frac{e \sin \omega}{\sin (v_1 + \omega)}$$

From this and the value of $\sin (v_1 + \omega)$, it now follows that:

$$e \sin \omega = \frac{2\sqrt{AB}}{A + B} \cdot \frac{Z_2 - Z_1}{Z_2 + Z_1} \qquad (171)$$

Now we can find e and ω.

At the time of periastron passage $v = 0°$ and $p' = K(1 + e) \cos \omega$. Because K, e, ω are already known, we can compute the ordinate length p'. Actually two points on the radial velocity curve have this length. But at A we have $v + \omega = 0°$, and at B we have $v + \omega = 180°$. Because ω is already known, this determines the periastron point on the radial velocity curve. Thus we know also the periastron passage time (T) on the time axis.

The expression for half the amplitude was:

$$K = \frac{n\,a \sin i}{\sqrt{(1 - e^2)}} = \frac{2\pi}{P}\,\frac{a \sin i}{\sqrt{(1 - e^2)}}$$

From this we find:

$$a \sin i = \frac{KP}{2\pi}\sqrt{(1 - e^2)} = \frac{(A + B)}{4\pi}\,P\sqrt{(1 - e^2)}$$

The unit of A, B and K is km/sec. If we take the day as unit for the period P, we get an additional factor $60 \times 60 \times 24 = 86{,}400$.

$$a \sin i = 13{,}751\,KP\sqrt{(1 - e^2)} \qquad (172)$$

Obviously $a \sin i$ is found in kilometers, where a is the semi-major axis of the absolute orbit of the brighter star. We get a similar expression for the fainter star. If we take a million kilometers as our unit for a we get:

$$a \sin i = 0.01375\,KP\sqrt{(1 - e^2)} \qquad (173)$$

In this method we have to measure V_0, A, B, Z_1, Z_2 and can find V_0, T, e, ω, $a \sin i$. We summarize the more important formulae:

$$\left.\begin{aligned} &V_r = K\{e\cos\omega + \cos(v+\omega)\}, \qquad K = \tfrac{1}{2}(A+B) \\ &e\cos\omega = \frac{A-B}{A+B}, \qquad e\sin\omega = \frac{2\sqrt{AB}}{A+B}\cdot\frac{Z_2-Z_1}{Z_2+Z_1} \\ &a\sin i = 13{,}751\, KP\sqrt{(1-e^2)} \end{aligned}\right\}$$

To compare the elements with the observations we compute the theoretical velocities from the elements with the help of the following expressions:

$$\left.\begin{aligned} &M_a = \frac{2\pi}{P}(t-T) = n(t-T) = E - e\sin E \\ &\tan\frac{v}{2} = \sqrt{\frac{1+e}{1-e}}\tan\frac{E}{2} \\ &V_r = V_0 + Ke\cos\omega + K\cos(v+\omega) \end{aligned}\right\}$$

The value of v for each M_a can be read from the Allegheny tables for $e < 0.77$. If now for example the slope of the branches of the radial velocity curve does not agree with the observations, this means that e needs a correction, as will ω and T. After a new computation the correction in V_0 is found as the mean residual. Better still are the differential corrections provided by F. Schlesinger.

57. *Method of K. Schwarzschild and W. Zurhellen.* We have found in the previous method that the radial velocity curve consisted of a constant part $Ke\cos\omega = \frac{1}{2}(A-B)$ and a periodic part $K\cos(v+\omega)$. We now introduce a new axis, which is shifted $Ke\cos\omega$ with respect to the V_0 axis. For obvious reasons we will call this new axis the median axis. On this axis the radial velocity curve has the equation (Figure 78):

$$V_r = K\cos(v+\omega) \tag{174}$$

In the maximum the velocity is $+ K$; in the minimum, $- K$; thus the median axis has to be drawn exactly between the maximum and the minimum values. As before K is half the amplitude. The formula shows that V_r plotted against $(\upsilon + \omega)$ gives a cosine curve. Thus V_r plotted against υ gives a shifted cosine

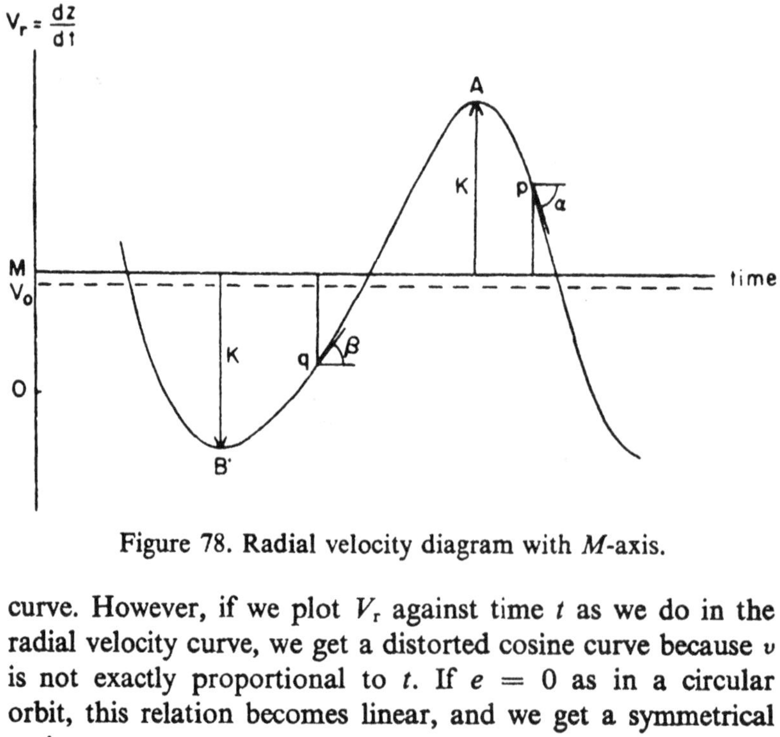

Figure 78. Radial velocity diagram with *M*-axis.

curve. However, if we plot V_r against time t as we do in the radial velocity curve, we get a distorted cosine curve because υ is not exactly proportional to t. If $e = 0$ as in a circular orbit, this relation becomes linear, and we get a symmetrical cosine curve.

For periastron and apastron we get now:

$$p = K \cos \omega, \qquad q = - K \cos \omega \qquad (175)$$

The ordinates thus have equal size but opposite sign, and are $\frac{1}{2}P$ apart in time. Usually we find two pairs of points, but only one pair suffices. The periastron point is located where the curve is steeper, the apastron where the slope of the curve is gentler. K. Schwarzschild used the following method (Figure 79): Trace the curve on transparent paper, shift it $\frac{1}{2}P$ and rotate it 180°. The intersections which are on different branches are the periastron and apastron. The periastron point is on the steeper branch. Now we immediately also know T. Further we know:

$$K, \qquad K\cos\omega = p, \qquad Ke\cos\omega = \tfrac{1}{2}(A - B)$$

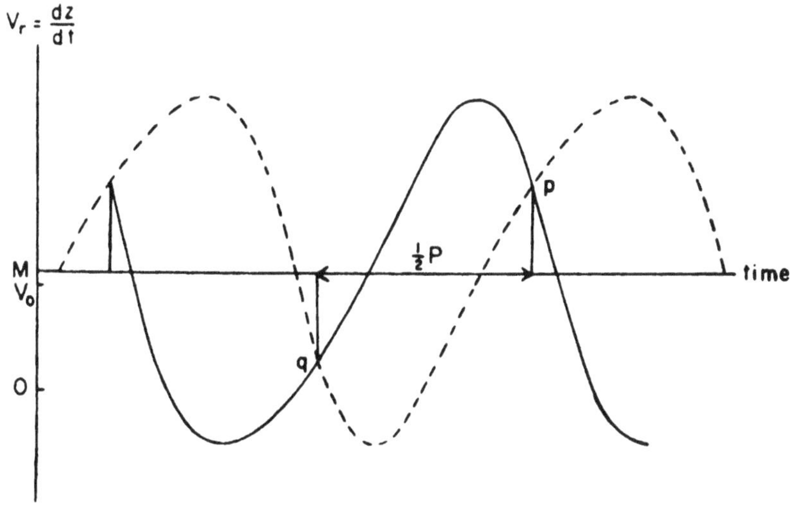

Figure 79. Determination of periastron point p and apastron point q.

For the last equation we must determine the V_0 axis. Now we find:

$$e = \frac{A - B}{2p}, \qquad \cos\omega = \frac{p}{K} = \frac{-q}{K} = \frac{p - q}{2K} \tag{176}$$

To determine the sign we know in addition that $\tan\alpha = dp/dt < 0$, if $\sin\omega > 0$. The $a \sin i$ we find as before. In this method we must measure K, V_0, M, p, dp/dt and find V_0, T, e, ω, $a \sin i$.

W. Zurhellen computed the change in the radial velocities. We will do this only for the points of periastron and apastron.

$$\left.\begin{aligned} \tan \alpha &= \frac{dp}{dt} = -K \sin \omega \left(\frac{dv}{dt}\right)_p \\ \tan \beta &= \frac{dq}{dt} = +K \sin \omega \left(\frac{dv}{dt}\right)_q \end{aligned}\right\} \quad \text{where} \left(\frac{dv}{dt}\right)_p > \left(\frac{dv}{dt}\right)_q$$

According to Kepler's second law $r^2\, dv/dt = h$ so that:

$$a^2 (1-e)^2 \left(\frac{dv}{dt}\right)_p = a^2 (1+e)^2 \left(\frac{dv}{dt}\right)_q$$

We then have:

$$\frac{\tan \alpha}{\tan \beta} = -\frac{(1+e)^2}{(1-e)^2} = -d^2, \qquad d = \sqrt{-\frac{\tan \alpha}{\tan \beta}}$$

$$e = \frac{d-1}{d+1} \tag{177}$$

This gives the eccentricity.

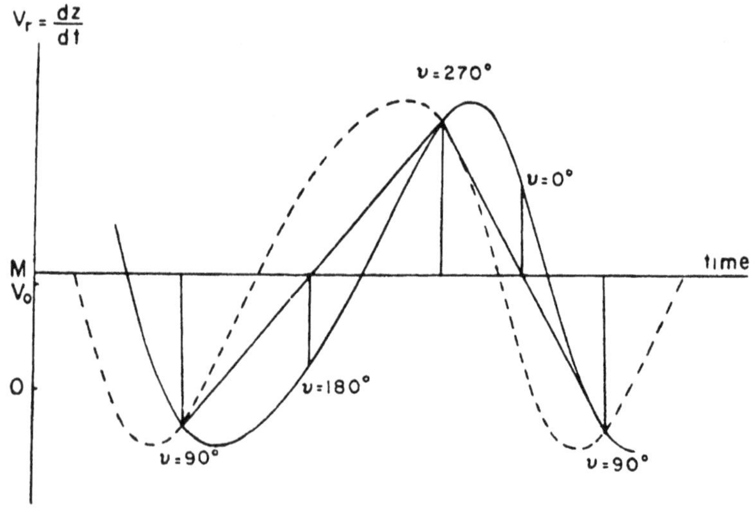

Figure 80. Determination of points where the true anomaly equals 90° and 270°.

We can now work with two points in the orbit where $v = 90°$ and $-90°$ respectively (Figure 80):

$$\left.\begin{aligned} V_{r1} &= K\cos(+90+\omega) = -K\sin\omega \\ V_{r2} &= K\cos(-90+\omega) = +K\sin\omega \end{aligned}\right\} \tag{178}$$

We have thus: $V_{r1} = -V_{r2}$, $E_1 = -E_2$, $M_1 = -M_2$, $t_1 - T = T - t_2$. These points are symmetrically placed with respect to the median axis and with respect to the time of periastron passage. Therefore, rotate the curve 180° and shift it till the times of periastron passage on both curves coincide. The intersections of both curves give the desired points where $v = \pm 90°$. The lines that connect these points intersect the time of periastron and apastron on the median axis.

$$\left.\begin{aligned} \sin\omega &= \frac{V_{r1}}{K} = -\frac{V_{r2}}{K} = \frac{V_{r1} - V_{r2}}{2K} \\ \cos\omega &= \frac{p}{K} = -\frac{q}{K} = \frac{p-q}{2K} \\ \tan\omega &= \frac{V_{r1} - V_{r2}}{p-q} \end{aligned}\right\} \tag{179}$$

From this ω follows. Now define in Figure 81:

$$E = E_1 = -E_2, \qquad e = \sin\varphi = \cos E$$

$$\left.\begin{aligned} M_1 &= \frac{2\pi}{P}(t_1 - T) = +E - \cos E \sin E \\ M_2 &= \frac{2\pi}{P}(t_2 - T) = -E + \cos E \sin E \end{aligned}\right\}$$

By subtraction it follows that:

$$M_1 - M_2 = \frac{2\pi}{P}(t_1 - t_2) = 2E - 2\cos E\sin E = 2E - \sin 2E$$

For a certain e one can find E and thus $M_1 - M_2$ and $t_1 - t_2$; therefore e can be tabulated against $t_1 - t_2$ or against the phase difference $(t_1 - t_2)/P$. For a circle the phase difference equals $\frac{1}{2}$.

$$D - \tfrac{1}{2} = \frac{t_1 - t_2}{P} - \tfrac{1}{2} \tag{180}$$

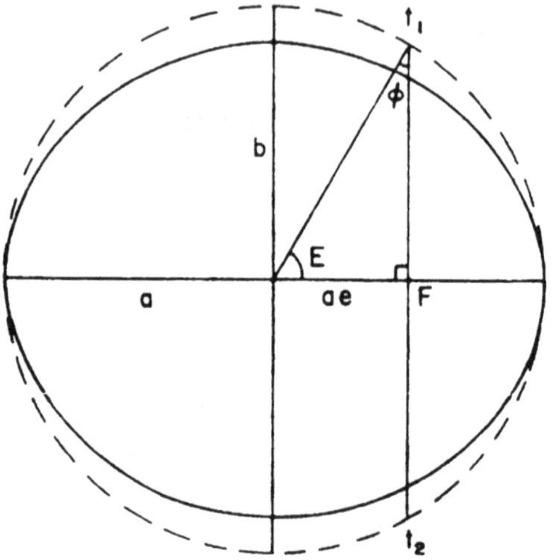

Figure 81. The eccentric anomaly in an ellipse.

We call this the phase shift. This shift is found to be linear with the eccentricity for $e < 0.5$. For small e, this is easily seen from Kepler's second law; from Figure 82 we can see that, to a first approximation:

$$D = \frac{t_1 - t_2}{P} = \frac{\frac{1}{2}\pi ab \pm ae.2b}{\pi ab} = \tfrac{1}{2} \pm \frac{2e}{\pi}$$

$$e = \frac{\pi}{2}\left| D - \tfrac{1}{2} \right| \tag{181}$$

A closer approximation gives $e = 1.63 \left| D - \frac{1}{2} \right|$. From this, e is found.

The differential corrections have been described by R. H. Curtiss.

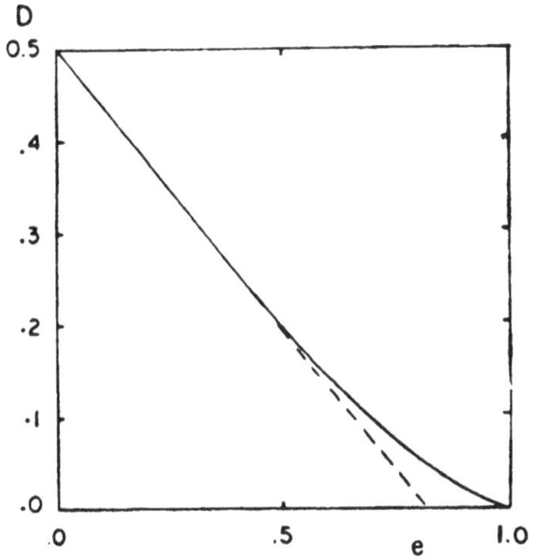

Figure 82. Relation between phase difference and eccentricity.

58. *Method of J. Wilsing and H. N. Russell.* This method is most useful for the case of small eccentricity and not at all advisable for high eccentricity. We have seen that the radial velocity curve (V_r against t) is a cosine curve in case of a circular orbit, but that the curve becomes somewhat distorted if the orbit is elliptical. The distortion will therefore be related to the eccentricity. Consequently we add a term containing twice the argument and expect that e somehow will enter into the coefficient.

$$V_r = A_0 + A_1 \cos (nt + \alpha_1) + A_2 \cos (2nt + \alpha_2) \qquad (182)$$

Because P is known, we know n, and in addition we know t for each normal point. Our first problem is to determine the five constants A_0, A_1, A_2, α_1, α_2. Five suitable points on the radial velocity curve will suffice. However, the constants will be determined even better by harmonic analysis (Fourier analysis) or by least squares, because then we use all points on the curve.

Our next problem is to find what functions these constants are of the usual elements.

$$z = r\sin(v + \omega)\sin i = r\sin i\{\sin v\cos\omega + \cos v\sin\omega\}$$

$$\frac{dz}{dt} = \sin i\left\{\cos\omega\frac{d}{dt}(r\sin v) + \sin\omega\frac{d}{dt}(r\cos v)\right\}$$

In the two body problem we have now the following series:

$$\left.\begin{aligned} r\sin v &= \phantom{\tfrac{3}{2}ae} + a\sin M_a + \tfrac{1}{2}ae\sin 2M_a + \dots \\ r\cos v &= \tfrac{3}{2}ae + a\cos M_a + \tfrac{1}{2}ae\cos 2M_a + \dots \end{aligned}\right\} \qquad (183)$$

We have if M_0 is the mean anomaly at time $t = 0$.

$$M_a = n(t - T), \quad M_0 = -nT, \quad M_a = M_0 + nt, \quad \frac{dM_a}{dt} = n$$

From this it follows that:

$$\frac{d}{dt}(r\sin v) = +na(\cos M_a + e\cos 2M_a)$$

$$\frac{d}{dt}(r\cos v) = -na(\sin M_a + e\sin 2M_a)$$

Thus we find finally:

$$\begin{aligned} \frac{dz}{dt} &= na\sin i(\cos\omega\cos M_a + e\cos\omega\cos 2M_a - \sin\omega\sin M_a \\ &\qquad - e\sin\omega\sin 2M_a) \\ &= na\sin i\{\cos(\omega + M_a) + e\cos(\omega + 2M_a)\} \\ &= na\sin i\{\cos(\omega + M_0 + nt) + e\cos(\omega + 2M_0 + 2nt)\} \end{aligned}$$

$$V_r = V_0 + \frac{dz}{dt} \qquad (184)$$

A comparison with the Fourier sequence gives:

$$\left.\begin{aligned} V_0 &= A_0 \\ na\sin i &= A_1, \qquad \omega + M_0 = \alpha_1 \\ nea\sin i &= A_2, \qquad \omega + 2M_0 = \alpha_2 \end{aligned}\right\} \qquad (185)$$

This we can write as:

$$\left.\begin{aligned} &V_0 = A_0 \\ &a \sin i = \frac{A_1}{n}, && M_0 = \alpha_2 - \alpha_1 = -nT \\ &e = \frac{A_2}{A_1}, && \omega = 2\alpha_1 - \alpha_2 \end{aligned}\right\} \quad (186)$$

If we take the P in days we have:

$$a \sin i = 86{,}400 \frac{A_1}{n} = 13{,}751\, A_1 P \quad (187)$$

From A_0, A_1, A_2, α_1, α_2 we find V_0, $a \sin i$, e, ω, T. This holds only when $e < 0.1$, or for almost circular orbits. Russell increased the value of this method by taking into account second order terms. For $e < 0.3$ we have with sufficient accuracy, when the new elements are given with a prime:

$$\begin{aligned} a' &= a\{1 + \tfrac{1}{2} e^2 (1 + \tfrac{1}{4} \cos 2\,\omega)\} \\ e' &= e\{1 + \tfrac{1}{4} e^2 (1 - \tfrac{1}{6} \cos 2\,\omega)\} \\ \omega' &= \omega - \tfrac{1}{6} e^2 \sin 2\,\omega \\ M_0' &= M_0 + \tfrac{1}{24} e^2 \sin 2\,\omega \end{aligned}$$

If one includes third order terms the computation becomes impractical. An advantage of this method is that we can use the normal points directly and do not have to draw the curve at all. It is not even necessary that the maximum and minimum parts of the curve be covered with observations.

The differential corrections have been analyzed by W. J. Luyten.

59. *Method of H. N. Russell.* We determine P according to the previously mentioned method. We know also $n = 2\pi/P$ and write the formula for the radial velocity curve as follows, where we take $G = V_0 + Ke \cos \omega$:

$$\begin{aligned} V_r &= V_0 + Ke \cos \omega + K \cos (v + \omega) \\ V_r &= G + K \cos (v + \omega) \end{aligned} \quad (188)$$

The maximum value is $G + K$, the minimum value is $G - K$. Both can be measured. Then G and K follow separately. For each normal point we have:

$$\cos(\upsilon + \omega) = \frac{V_r - G}{K} \tag{189}$$

For certain V_r we can thus find the corresponding $(\upsilon + \omega)$.

$M_a = n(t - T)$. Let $M_p = nT$, then $M_a + M_p = nt$ is known. The M_p is here the mean anomaly at the time of periastron T, when thus $M_a = 0$. For each normal point we know therefore:

$$(\upsilon + \omega) - (M_a + M_p) = (\upsilon - M_a) + (\omega - M_p) \tag{190}$$

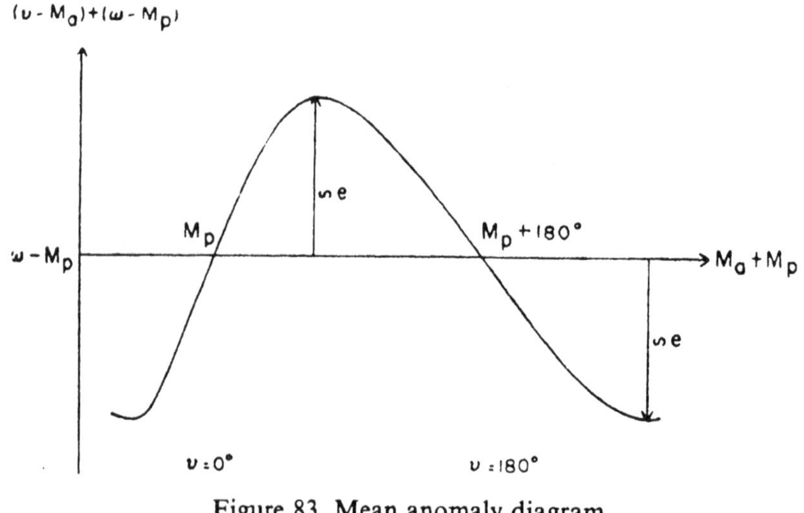

Figure 83. Mean anomaly diagram.

We plot this against $(M_a + M_p)$ and call this the anomaly diagram (Figure 83). Both coordinates are thus in degrees. The quantity $(\omega - M_p)$ is a constant while $(\upsilon - M_a)$ varies between equal positive and negative limits, which are almost a linear function of e. Draw the median line in the anomaly diagram. Half the amplitude gives e with help of the following table. See also Figure 84.

e	0.1	0.2	0.3	0.4	0.5	0.6	0.7	0.8	0.9
ampl.	11°.5	23°.0	34°.8	46°.8	59°.2	72°.3	86°.4	102°.3	122°.2

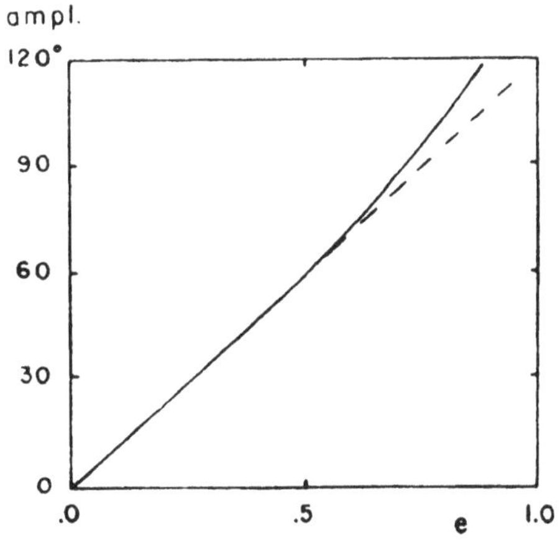

Figure 84. Relation between amplitude of mean anomaly curve and eccentricity.

From K and e we now find $a \sin i$. The distance from the median line to the zero point gives us $(\omega - M_p)$. The intersections of the curve with the median line correspond to periastron and apastron because $v = M_a$ for these points. Again periastron is situated on the branch with the largest slope. The x values are then M_p and $M_p + 180°$. From this we find M_p and thus T. From $(\omega - M_p)$ we find ω. A change in M_p produces a horizontal shift; a change in $(\omega - M_p)$ gives a vertical shift.

In this method we measured $G + K$, $G - K$, $\omega - M_p$, M_p and found the elements e, $a \sin i$, T, ω.

60. *Method of K. Laves and W. F. King.* This method makes clear how the radial velocity curve is built up and has to be of the form:

$$V_r = V_0 + Ke \cos \omega + K \cos (\upsilon + \omega)$$

V_0 is the radial velocity of the gravity center, which is the projection of the space motion onto the line of sight. We will see that $Ke \cos \omega$ is the projection of the constant velocity component which is perpendicular to the major axis in space. Further $K \cos (\upsilon + \omega)$ is the projection of the velocity component which is constant in size but variable in direction, namely always perpendicular to the radius vector in space.

The method makes use of the so-called hodograph. We have

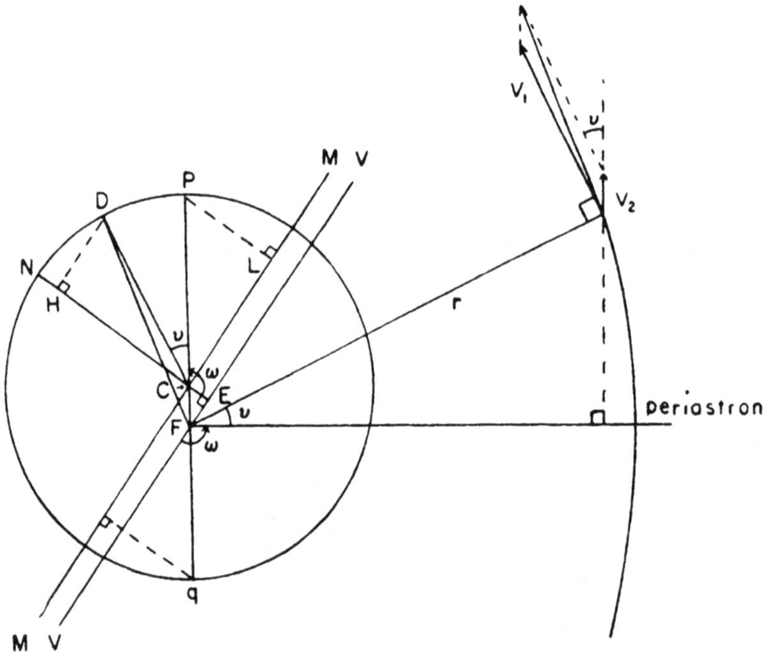

Figure 85. The hodograph of the elliptical orbit.

seen that every position ($r \cos v$, $r \sin v$) in the true orbit has a variable velocity (Figure 85). The components can be taken as a constant speed $V_1 e$ perpendicular to the major axis, and a constant speed V_1 with variable direction but always perpendicular to the radius vector. In the hodograph they are represented by FC and CD respectively while FD represents the instantaneous speed. We now have to project these velocities upon the line of sight. We draw the line of the nodes. The projection of the line of sight on the true plane will be EN perpendicular to the line of the nodes. The projections of the velocities upon the line EN will be EC and CH respectively.

We know a priori that we are unable to determine the inclination between the orbital plane and the tangential plane and we may as well assume $i = 90°$, that is that the line of sight is constantly contained in the plane of motion. This permits us to avoid the elliptical projection of the hodographic circle on the plane of the sky which passes through the line of the nodes. However, the radius of the hodographic circle becomes smaller and equal to K.

In the radial velocity curve the V_0 and M axes have to be drawn. We find K as before as half the amplitude of the radial velocity curve and construct a circle with K as radius. The center of the circle falls on the M axis. Divide the period of the radial velocity into a whole number of equal intervals and project the corresponding radial velocities onto the circle. Around periastron these are widely spaced and around apastron very compactly distributed. This determines the periastron point which can be checked by the Schwarzschild method. We have now (Figure 86):

$$K = \tfrac{1}{2}(A + B), \quad \text{CF} = Ke, \quad \text{pL} = K \cos \omega = p$$

For the two projections on the line EN we get:

$$\text{EC} = Ke \cos \omega, \qquad \text{CH} = K \cos (v + \omega) \qquad (191)$$

At F draw the perpendicular line $s = FG$. Then we have:

$$s^2 = K(1-e)\,K(1+e) = K^2(1-e^2), \quad s = K\sqrt{(1-e^2)}$$

We find with help of this equation:

$$a \sin i = 13{,}751\, KP \sqrt{(1-e^2)} = 13{,}751\, Ps \qquad (192)$$

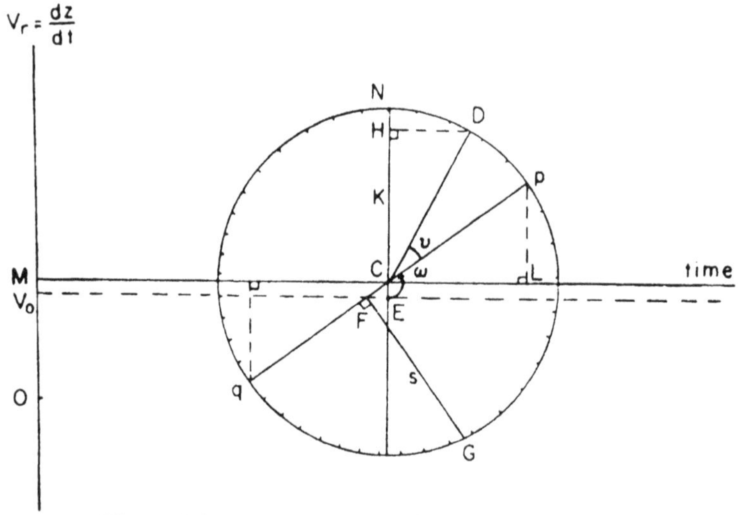

Figure 86. Determination of periastron point p.

The numerical factor is included because s is in kilometers per second, and we like to have P in days. In this method we measured K, p, V_0, M, s and find $T, V_0, \omega, e, a \sin i$.

61. *Mass function* (*one spectrum visible*). We have $\mathfrak{M}_1 a_1 = \mathfrak{M}_2 a_2$ for the mass balance, from which follows:

$$\frac{\mathfrak{M}_2}{\mathfrak{M}_1} = \frac{a_1}{a_2} = \frac{a_1 \sin i}{a_2 \sin i}, \qquad B_2 = \frac{\mathfrak{M}_2}{\mathfrak{M}_1 + \mathfrak{M}_2} = \frac{a_1}{a_1 + a_2} = \frac{a_1}{a} \qquad (193)$$

If we take as units the mass of the sun, the astronomical unit and the year, Kepler's third law can be written:

$$\mathfrak{M}_1 + \mathfrak{M}_2 = \frac{a^3}{P^2}$$

Thus we have:

$$\left(\frac{\mathfrak{M}_2}{\mathfrak{M}_1+\mathfrak{M}_2}\right)^3=\frac{a_1^3}{a^3}=\frac{a_1^3}{(\mathfrak{M}_1+\mathfrak{M}_2)\,P^2}$$

$$\frac{\mathfrak{M}_2^3}{(\mathfrak{M}_1+\mathfrak{M}_2)^2}=\frac{a_1^3}{P^2} \qquad (194)$$

We now take a million kilometers as our unit for a and the P in days. This gives:

$$\frac{a^3}{149.5^3}\,\frac{365.25^2}{P^2}=\frac{a^3}{25.0\,P^2}$$

We have still to multiply by $\sin^3 i$ because only $a \sin i$ can be observed. The following mass function is therefore the only thing we can find out about the mass:

$$f(\mathfrak{M})=\frac{(\mathfrak{M}_2 \sin i)^3}{(\mathfrak{M}_1+\mathfrak{M}_2)^2}=\frac{(a_1 \sin i)^3}{25\,P^2} \qquad (195)$$

Kepler's third law expressed in these units gives:

$$\mathfrak{M}_1+\mathfrak{M}_2=\frac{a^3}{25\,P^2} \qquad (196)$$

62. *Rotation effect.* Stars also rotate just as does the sun. Because the shape of the absorption lines changes when the rotation axis is close to the plane of the sky the rotation of single stars can be studied. During the eclipse of a double star we can discover something about it. Just before minimum only a crescent of the eclipsed star appears which has a velocity of recession while just after the minimum that crescent appears which has a rotation speed directed towards us. Outside the eclipse the halves of the stars balance each other and give a broadening of the absorption lines.

Some spectroscopic double stars with i about 90° can also be observed as eclipsing stars (Figure 87). At the time of mid-eclipse both stars are in the line of sight, and for circular orbits

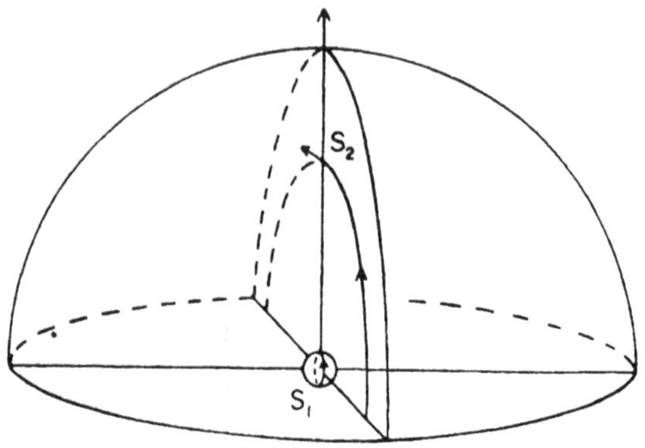

Figure 87. The observed rotation effect shows that in most cases the rotation of the stars is in the same direction as the revolution.

their motions are perpendicular to it. This corresponds to the positions E and D in the radial velocity diagram (Figure 88). We consider the absolute orbit of the bright star with respect to the center of gravity. The eclipse of the bright star occurs at E on the decreasing branch of the radial velocity curve where it intersects the V_0 axis.

For an elliptical orbit this is not strictly true. For $i = 90°$, at mid-eclipse the radius vector coincides with the line of sight and the speed V_1 is perpendicular to it. However, the component V_1e is perpendicular to the major axis and has thus a component in the line of sight. For all practical purposes e is small and in general the mid-eclipse will occur very close to E.

The rotation of the stars is usually in the same direction as the orbital motion while the rotation axis probably is practically perpendicular to the orbital plane. In other words for such a star the axis is practically in the plane of the sky, and the direction of rotation of the limb is practically in the direction of the line of sight. We observe $V \sin i$ if $i \neq 90°$. Just before the primary minimum the crescent of the bright star is therefore going away from us giving a positive rotation radial velocity. If the eclipse is total the bright star cannot be seen during minimum and no rotation effect is observed. Just after primary minimum the velocity of rotation is negative. The observations confirm this.

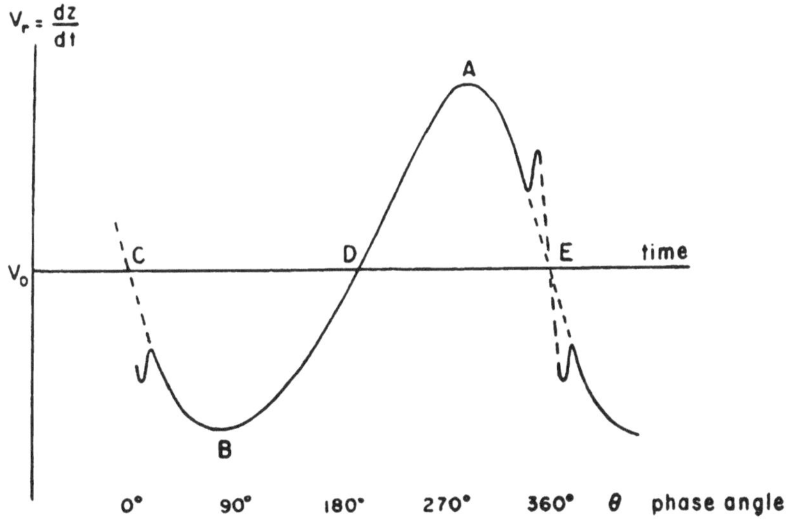

Figure 88. The influence of the rotation effect. The primary eclipse happens at C and E.

63. *Gas streams.* A statistical study of spectroscopic double stars showed that there was a systematic tendency for the ω's to be in the first quadrant. This would indicate the earth to be in a preferred position with respect to these stars, and this is something which no astronomer is willing to accept. The conclusion is that the real or physical situation of the double star must be somewhat more complicated than the simple geometrical picture we used in the derivation of the spectroscopic elements.

O. Struve studied a spectroscopic double which was also an

eclipsing variable. First from the light curve he determined the orbital elements. Using these elements he computed the expected radial velocity curve. He found that the largest difference between the computed and observed curves occurred just before primary eclipse (Figure 89). This can be explained by a gas stream around the binaries, something like very exaggerated prominence activity. The binaries in which this effect has been found are all very close so that one can expect interactions like tidal waves or more violent phenomena.

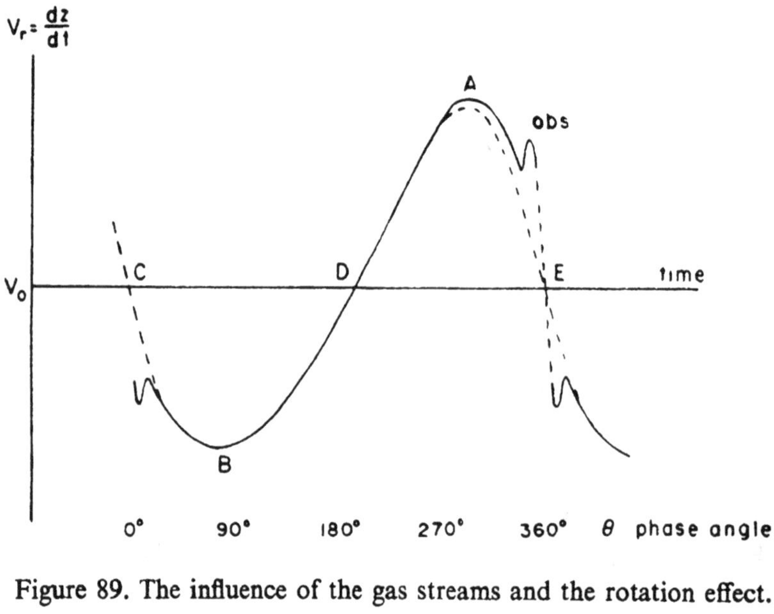

Figure 89. The influence of the gas streams and the rotation effect.

Figure 90 illustrates such a system of streams. Let S_1 be the brighter and more massive star of which the radial velocity curve has been observed. The speed of S_2 with respect to the gravity center will be larger than that of S_1, both stars having the same period *P*. Just before the eclipse by the S_2 star begins a stream of gas is projected upon the apparent disk of star S_1. The radial velocity of this stream is one of recession, and larger than the radial component of the orbital motion of S_1. Just after the eclipse an approaching stream of gas is observed in front of

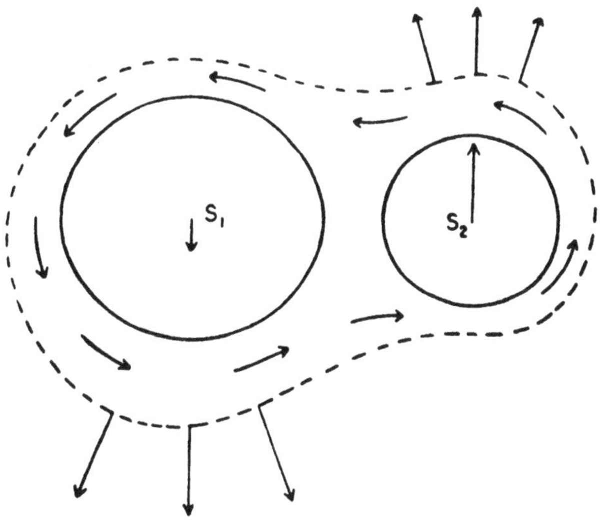

Figure 90. Gaseous streams in close binary systems.

star S_1. If the absorption lines originating in the reversing layer of the S_1 star and in the gas mass are blended the result will be a distortion of the radial velocity curve of this brighter star, in the sense that we have a positive amount of distortion just before eclipse and a negative amount just after eclipse. The conditions of ionization of the two sides of the stream are not the same and the blending may not produce equal distortions of the radial velocity curve on both sides of the time of eclipse. Without applying the proper correction we may thus find incorrect orbital elements for close doubles.

64. *Two spectra visible.* When the components have comparable brightnesses, the spectral lines will be seen periodically doubled (Plate VII). In this way the first spectroscopic double was discovered. We then observe two radial velocity curves, which intersect on the V_0 axis. At those times both components have a speed in space perpendicular to the line of sight and parallel with the line of the nodes through the center of gravity. This is thus caused by the similarity of both absolute ellipses with respect to the center of gravity. We have thus one V_0 axis, but two M axes because each curve has a median axis. For the same time the V_0, i, e are the same for both components. Also the v's are the same, but the ω's differ by 180°. We can now prove that for any given time the radial velocities relative to the V_0 axis have the same proportionality factor. The curves are similar with respect to the V_0 axis, because the absolute orbits are

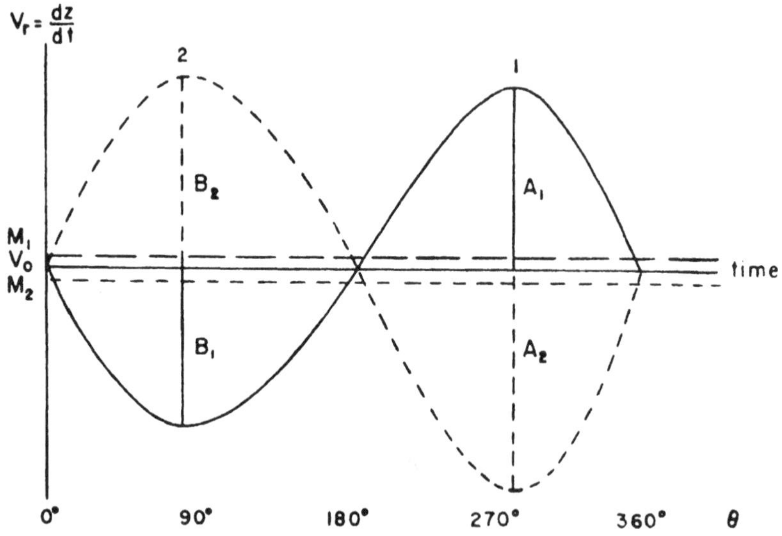

Figure 91. Radial velocity diagrams for the case of two spectra visible.

similar with respect to the center of gravity. Observations on both curves very close to the V_0 axis influence each other, and should be corrected for the blending effect, or omitted. The spectral line is then broadened (Figure 91).

Let subscript 1 belong to the brighter component, which has the stronger and therefore better visible absorption lines.

$$V_{r1} = K_1 \{e \cos \omega + \cos (v + \omega)\}$$
$$V_{r2} = K_2 \{e \cos (\omega + 180) + \cos (v + \omega + 180)\}$$
$$= -K_2 \{e \cos \omega + \cos (v + \omega)\}$$
$$\frac{V_{r1}}{V_{r2}} = -\frac{K_1}{K_2} \qquad (197)$$

Using each of the curves we can derive the orbital elements as before. However, we can also utilize the relative curve which corresponds to the relative ellipse. For this case we need to measure only the relative distance between two absorption lines; therefore we do not need the comparison spectra.

$$V_{rel} = V_{r1} - V_{r2} = K_{rel} \{e \cos \omega + \cos (v + \omega)\} \qquad (198)$$
$$K_{rel} = K_1 + K_2 = \frac{n(a_1 + a_2) \sin i}{\sqrt{(1 - e^2)}} = \frac{n\, a \sin i}{\sqrt{(1 - e^2)}} \qquad (199)$$

Here we find thus the semi-major axis of the relative orbit.

$$a \sin i = \frac{(K_1 + K_2)}{n} \sqrt{(1 - e^2)} = (K_1 + K_2) \frac{P}{2\pi} \sqrt{(1 - e^2)}$$

When we express a in kilometers and P in days we get:

$$a \sin i = 13{,}751\, (K_1 + K_2) P \sqrt{(1 - e^2)} \qquad (200)$$

When we take a million kilometers as our unit for a and P in days:

$$a \sin i = 0.01375\, (K_1 + K_2) P \sqrt{(1 - e^2)} \qquad (201)$$

Figure 92 shows the radial velocity curves of a typical W Ursae Majoris system. The orbit is circular and the velocity curves are thus sine curves. The unusual fact is that the primary eclipse often occurs at the middle of the increasing branch of radial velocity curve 1.

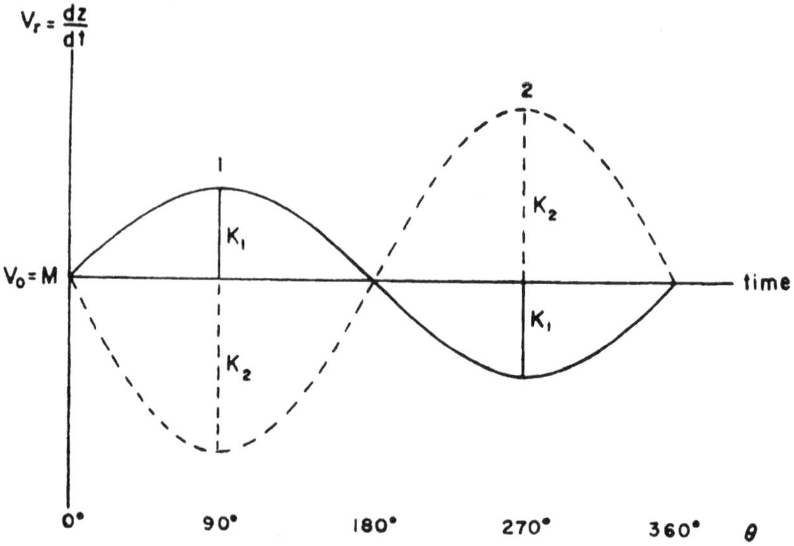

Figure 92. Radial velocity diagrams for a typical W Ursae Majoris variable.

65. *Mass ratio* (*two spectra*). The mass ratio is the reciprocal of the ratio of the amplitudes.

$$\frac{\mathfrak{M}_2}{\mathfrak{M}_1} = \frac{a_1}{a_2} = \frac{a_1 \sin i}{a_2 \sin i} = \frac{K_1}{K_2} = \frac{A_1 + B_1}{A_2 + B_2} = \frac{A_1}{A_2} = \frac{B_1}{B_2}$$

because for the extremes:

$$\frac{A_1}{A_2} = \frac{K_1(1 + e \cos \omega)}{K_2(1 + e \cos \omega)} = \frac{K_1}{K_2}$$

As a rule we find that the brighter component is the more massive one, and has the smaller amplitude. The K_1 and K_2 are positive.

$$\frac{\mathfrak{M}_2}{\mathfrak{M}_1} = \frac{K_1}{K_2} \tag{202}$$

Further we have when a is in astronomical units and P in years:

$$\mathfrak{M}_1 + \mathfrak{M}_2 = \frac{a^3}{P^2}, \qquad (\mathfrak{M}_1 + \mathfrak{M}_2)\sin^3 i = \frac{(a \sin i)^3}{P^2}$$

From the mass balance and a property of ratios we have in addition:

$$\mathfrak{M}_1 = \frac{a_2}{a}(\mathfrak{M}_1 + \mathfrak{M}_2) = \frac{a_2}{a}\frac{a^3}{P^2} = \frac{a_2 a^2}{P^2}$$

If again we take a million kilometers as our unit for a and P in days we have:

$$(\mathfrak{M}_1 + \mathfrak{M}_2)\sin^3 i = \frac{(a \sin i)^3}{25\, P^2} \tag{203}$$

$$\left.\begin{aligned} \mathfrak{M}_1 \sin^3 i &= \frac{(a_2 \sin i)(a \sin i)^2}{25\, P^2} \\ \mathfrak{M}_2 \sin^3 i &= \frac{(a_1 \sin i)(a \sin i)^2}{25\, P^2} \end{aligned}\right\} \tag{204}$$

We can thus find the ratio of the masses, but the masses themselves or the sum of the masses only in combination with $\sin^3 i$. If the double star is in addition an eclipsing variable we can find the i from the light curve and then find the masses separately.

Statistically we can still make an estimate by considering all spectroscopic double stars. With respect to the random inclinations of the orbits the mean value of $\sin^3 i$ is 3/5. When the greater probability of discovery of systems of larger inclinations has been taken into account we get:

$$(\mathfrak{M}_1 + \mathfrak{M}_2)\sin^3 i = \tfrac{2}{3}\, n\, \overline{\mathfrak{M}} \tag{205}$$

where n is the number of systems and $\overline{\mathfrak{M}}$ is the averaged value of the total mass.

66. *Influence of reflection effect.* When both spectra are visible, the components have comparable luminosities. The reflection effect (chapter VI, paragraph 117) should be taken into account

in the determination of the masses. It causes the facing hemispheres to be somewhat brighter than the opposed ones. The centers of light of the two disks are not projected precisely on the centers of mass but are displaced toward the center of gravity of the system. The amplitudes measured correspond to radial velocities in an effectively smaller orbit. In other words, to the result for a or $a \sin i$ we must apply a correction factor somewhat greater than unity. This correction has been tabulated by G. P. Kuiper as a function of the relative sizes of the components and the spectral types.

Applied to a or $a \sin i$ the correction is not of great importance, but it does enter to the third power in the $\mathfrak{M} \sin^3 i$ as can be seen from formula (204). This will consequently produce appreciable changes in the computed masses. If the temperatures of the components are appreciably different the corrections will be different and the mass ratio will also be changed. However, in practice the mass ratio will change only slightly because the components are not very different if both spectra are visible.

If only a single spectrum is visible, the magnitude difference between the components is large. Therefore, in general the fainter component does not produce a measurable reflection effect upon the brighter one. It is the spectrum of the latter which is studied, and the correction to $f(\mathfrak{M})$ can be neglected.

67. *Ratio of intensities*. We are considering here the case of two spectra. The method can be applied to absorption line profiles obtained from high dispersion spectra. The reduction curve of the photographic plate can be found as before. We start then with the contour of the absorption line, or the intensity as a function of the wavelength in Angstrom units.

R. M. Petrie has given the formulae for the general case. However, this problem cannot be solved and one has to make some sort of assumption. Then we can derive the intensity ratio of both components or the magnitude difference between the two stars (Figure 93). We define:

x = coordinate in direction of the dispersion from the apparent center of the more intense line and always positive toward the weaker component.

$i(x)$ = intensity in spectral line

$I(x)$ = intensity of continuum

$f(x)$ = line profile undistorted by the light of other component, thus $f(x) = i(x)/I(x)$.

$R(x)$ = observed line profile when centers coincide

$r(x)$ = observed line profile when centers do not coincide

l = luminosity of fainter component in terms of that of the brighter one, thus $l = L_2/L_1 = I_2(x)/I_1(x)$.

We consider first the case when both centers coincide. The intensity at a point x in the spectral line is $i_1(x) + i_2(x)$, while the continuum by definition is $I_1(x) + I_2(x)$. The observed profile for the single line is thus:

$$R(x) = \frac{i_1(x) + i_2(x)}{I_1(x) + I_2(x)} = \frac{f_1(x) I_1(x) + f_2(x) I_2(x)}{I_1(x) + I_2(x)}$$

$$R(x) = \frac{f_1(x) + l f_2(x)}{1 + l} \tag{206}$$

We will make now the simplest assumption possible, namely that the line profiles are identical in form in the two stars. This assumption is reasonable because the stars are not very different if both spectra are visible. We assume thus $f_1(x) = f_2(x) = f(x)$. The formula becomes then $R(x) = f(x)$ or the observed line profile is identical with the original ones. Especially we have:

$$R(0) = f(0), \qquad R(\Delta) = f(\Delta) \tag{207}$$

We consider now the case when both centers do not coincide and the absorption lines in general are blended. The light at a point x in the spectral line is $i_1(x) + i_2(x - \Delta)$, and that in the continuum is $I_1\,(x) + I_2\,(x - \Delta)$. Here Δ is the difference in wavelengths between the two centers.

Proceeding as before we find:

$$r(x) = \frac{f_1(x) + l f_2(x - \Delta)}{1 + l} = \frac{f(x) + l f(x - \Delta)}{1 + l} \qquad (208)$$

Especially we have for $x = 0$ and $x = \Delta$:

$$r(0) = \frac{f(0) + l f(\Delta)}{1 + l} = \frac{R(0) + l\,R(\Delta)}{1 + l} \qquad (209)$$

$$r(\Delta) = \frac{f(\Delta) + l f(0)}{1 + l} = \frac{R(\Delta) + l\,R(0)}{1 + l} \qquad (210)$$

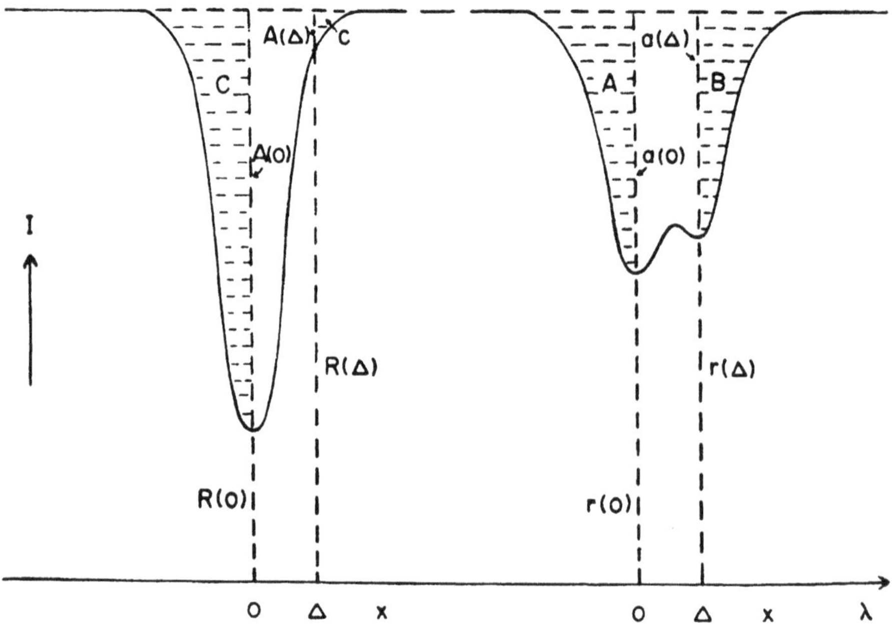

Figure 93. The two absorption lines are superimposed at the left and blended at the right.

We can solve for l from the first of these two formulae.

$$r(0) + l\,r(0) = R(0) + l\,R(\Delta)$$

$$l = \frac{r(0) - R(0)}{R(\Delta) - r(0)} = \frac{1 - R(0)/r(0)}{R(\Delta)/r(0) - 1} \tag{211}$$

Radial velocity data for both components must be available to supply a value of Δ, unless the latter is so great that $R(\Delta)$ may be taken as unity. In practice there is a better formula because the ratios employed tend to diminish the effect of errors of measurement. We must then divide the two formulae.

$$\frac{r(0)}{r(\Delta)} = \frac{R(0) + l\,R(\Delta)}{R(\Delta) + l\,R(0)}$$

$$r(0)R(\Delta) + l\,r(0)R(0) = r(\Delta)R(0) + l\,r(\Delta)R(\Delta)$$

$$l = \frac{r(0)R(\Delta) - r(\Delta)R(0)}{r(\Delta)R(\Delta) - r(0)R(0)} = \frac{r(0)/r(\Delta) - R(0)/R(\Delta)}{1 - r(0)R(0)/r(\Delta)\,R(\Delta)} \tag{212}$$

All quantities in the right member are known.

We can also introduce the absorptions of the absorption lines.

$$A(x) = 1 - R(x), \qquad a(x) = 1 - r(x) \tag{213}$$

For $x = 0$ and $x = \Delta$ we get respectively for coinciding centers:

$$A(0) = 1 - R(0) = 1 - f(0)$$
$$A(\Delta) = 1 - R(\Delta) = 1 - f(\Delta)$$

For centers not coinciding we find:

$$a(0) = 1 - r(0) = \frac{A(0) + l\,A(\Delta)}{1 + l}$$

$$a(\Delta) = 1 - r(\Delta) = \frac{A(\Delta) + l\,A(0)}{1 + l}$$

Proceeding as before we have:

$$\frac{a(\Delta)}{a(0)} = \frac{A(\Delta) + l\,A(0)}{A(0) + l\,A(\Delta)}$$

$$a(\Delta)A(0) + l\,a(\Delta)A(\Delta) = a(0)A(\Delta) + l\,a(0)A(0)$$

$$l = \frac{a(\Delta)A(0) - a(0)A(\Delta)}{a(0)A(0) - a(\Delta)A(\Delta)} = \frac{a(\Delta)/a(0) - A(\Delta)/A(0)}{1 - a(\Delta)\,A(\Delta)/a(0)\,A(0)} \tag{214}$$

When the lines are completely separated $A(\Delta) = 0$ and we find:

$$l = \frac{L_2}{L_1} = \frac{a(\Delta)}{a(0)} \tag{215}$$

The light ratio of the components is found then as the ratio of the line depths.

We will now determine l from measurements at certain points of the line profile from which we can find the areas of the absorption lines. Because more points are used this will be more accurate in practice. In addition, the distortion of the line profile due to finite resolving power of the spectrograph affects the line depths much more than the areas. Let us define:

A = integrated absorption in double line from the center of the more intense component, outward and away from weaker line.

B = integrated absorption in double line from center of weaker component outward and away from the more intense component.

C = one half the integrated absorption in the single line.

c = integrated absorption in the single line, outward, from Δ, to the point where the line wing merges with continuum.

The integrals of the absorptions are thus:

$$A = \int_{-\infty}^{0} \{1 - r(x)\}\,dx, \qquad B = \int_{\Delta}^{\infty} \{1 - r(x)\}\,dx$$

$$C = \int_{0}^{\infty} \{1 - R(x)\}\,dx = \int_{\Delta}^{\infty} \{1 - R(x - \Delta)\}\,dx$$

$$c = \int_{\Delta}^{\infty} \{1 - R(x)\}\,dx = \int_{-\infty}^{0} \{1 - R(x - \Delta)\}\,dx$$

The lines produced separately in each star have as integrated absorption:

$$F_1(0) = \int_{-\infty}^{0} \{1 - f_1(x)\}\, dx, \quad F_1(\Delta) = \int_{\Delta}^{\infty} \{1 - f_1(x)\}\, dx$$

$$F_2(0) = \int_{0}^{\infty} \{1 - f_2(x)\}\, dx = \int_{\Delta}^{\infty} \{1 - f_2(x - \Delta)\}\, dx$$

$$F_2(\Delta) = \int_{\Delta}^{\infty} \{1 - f_2(x)\}\, dx = \int_{-\infty}^{0} \{1 - f_2(x - \Delta)\}\, dx$$

Now forming $1 - R(x)$ and $1 - r(x)$ and integrating we can write using the assumption $F_1(x) = F_2(x) = F(x)$:

$$A = \frac{F_1(0) + l F_2(\Delta)}{1 + l} = \frac{F(0) + l F(\Delta)}{1 + l} \tag{216}$$

$$B = \frac{F_1(\Delta) + l F_2(0)}{1 + l} = \frac{F(\Delta) + l F(0)}{1 + l} \tag{217}$$

$$C = \frac{F_1(0) + l F_2(0)}{1 + l} = F(0) \tag{218}$$

$$c = \frac{F_1(\Delta) + l F_2(\Delta)}{1 + l} = F(\Delta) \tag{219}$$

Proceeding as before we have:

$$\frac{B}{A} = \frac{F(\Delta) + l F(0)}{F(0) + l F(\Delta)} = \frac{c + l C}{C + l c}$$

$$BC + l Bc = Ac + l AC$$

$$l = \frac{BC - Ac}{AC - Bc} = \frac{B/A - c/C}{1 - B c/A C} \tag{220}$$

This gives the light ratio in terms of measurable features of the observed single and double line profiles. For complete separation of the absorption lines we have $c = 0$.

$$l = \frac{L_2}{L_1} = \frac{B}{A} \tag{221}$$

The light ratio is then found from the ratio of the equivalent widths.

Petrie has also studied the case in which the line profiles are dissimilar in the two stars. He gives a discussion of the sources of error. The blending effect brings the apparent centers of the absorption lines components closer together. The lines at single phase may be broadened, due to finite duration of exposure and failure to secure spectra when the radial velocities of the components are identical. The error caused by distortion of the line profiles due to the finite resolving power of the spectrograph is eliminated in the last method using the areas of the absorption lines.

68. *Diameter of Cepheids.* The radial velocity curve of a Cepheid is almost the mirror image of the light curve. We will draw it here with $-V_r$ directed upward in order to compare it better with the light curve. Historically Cepheids were first considered as double stars, but the orbital elements determined from the radial velocity curve gave results which were highly improbable. The ω's are situated in a very small interval, the e's are larger than normal for stars of such short period, $a \sin i$ and $f(\mathfrak{M})$ are very small, V_0 is variable in some cases. Also the radial velocity curve itself and therefore the elements are often of a variable nature. Cepheids are known to be giants, much larger than the orbit computed under this assumption. This hypothesis was therefore rejected.

The pulsating nature of these stars based on observational evidence was first suggested by H. Shapley and worked out theoretically by A. S. Eddington. In a simple pulsation all parts would move outward and later inward in unison. According to this idea the star should be hottest and therefore brightest when it is most contracted.

Actually the star reaches light maximum at a later time when its spectrum lines are farthest displaced toward the violet, when the gases which cause the lines are moving outward fastest. The original pulsation theory thus is too simple a proposal. The currently accepted pulsation theory by M. Schwarzschild is more flexible and more complicated. The star's interior still pulsates in unison as before and sends shockwaves outwards reaching the higher levels later than the lower levels.

We thus know now that a Cepheid is a spherical star which changes its diameter. The disk area changes consequently during the period. However, the color and spectral type are also observed to change during the period; hence the temperature and the surface brightness change periodically. For a black body there is a linear relation between surface brightness and color, but the star may deviate from a black body model. Moreover, a Cepheid is a variable and thus not a normal star.

It can be assumed that there is a single valued functional relation between surface brightness and color. With this assumption we will find also such a linear relation from the observations, but with a slope different from that corresponding to black body radiation. The observational material is most complete in the case of δ Cephei. These observations must be of the highest possible accuracy. We will follow here the discussion by A. J. Wesselink (Figure 94).

The light curves of m_b and m_y give us then the color curve $C = m_b - m_y$. The radial velocity curve enables us to find the displacement curve (D) by integration. The best way is to write V_r as a sum of Fourier terms which can be integrated one by one producing also the probable errors. The displacement is positive for increasing size. For the unit we can conveniently take a million kilometers (Figure 95).

$$D_r = pD = p \int V_r \, dt = p \int \frac{dz}{dt} \, dt \qquad (222)$$

Here D_r is the real displacement in the radial direction in

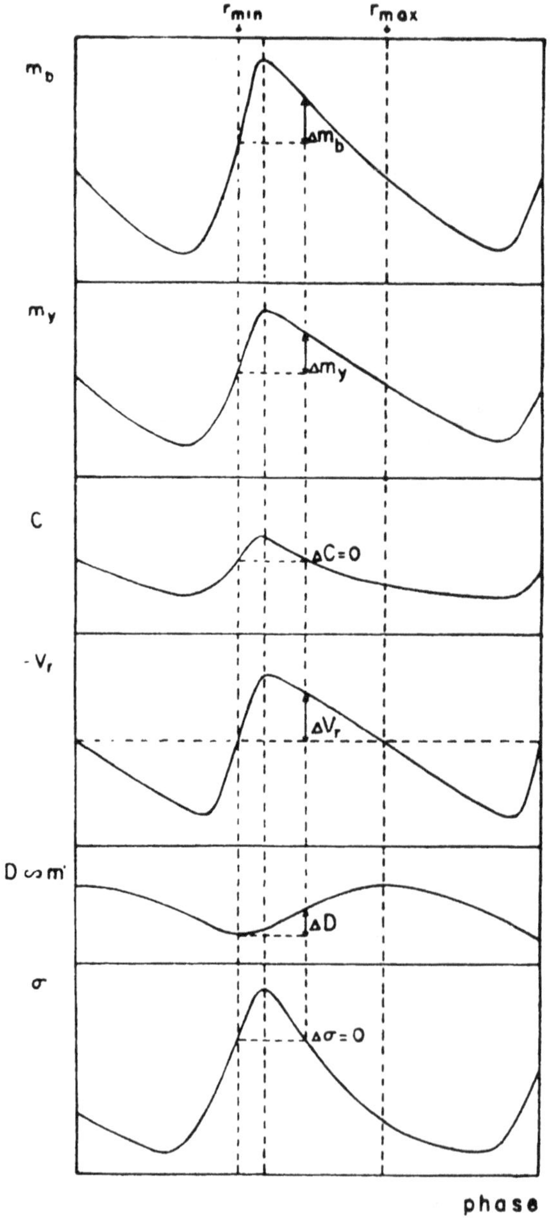

Figure 94. From top to bottom are given blue magnitude, yellow magnitude, color, negative radial velocity, displacement and surface brightness of δ Cephei.

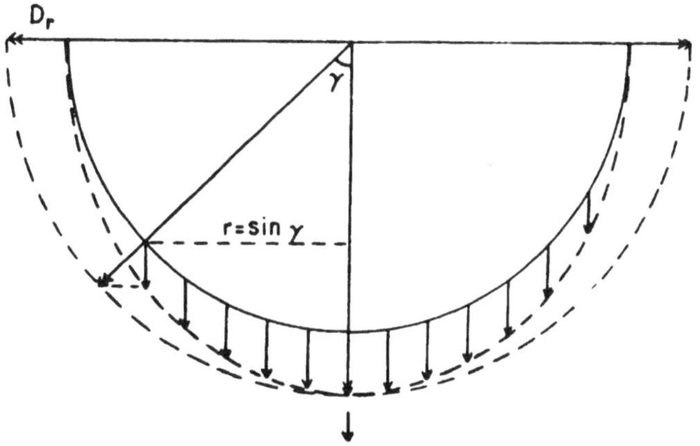

Figure 95. Relation between the pulsation in radial direction and the observed radial velocity in the line of sight.

space. In the line of sight towards the center of the sphere we indeed observe this quantity. However, in observing other points of the disk we observe only the component of D_r in the line of sight, namely $D_r \cos\gamma$, where γ is the angle between the radius and the line of sight. The observed displacement D is the averaged displacement of all components in the line of sight. The D is thus smaller than D_r so that the factor $p > 1$. For a uniformly luminous disk $p = 3/2$; for total darkening of the limb $p = 4/3$. The ratio of both extreme cases is 9/8 and so near unity that an uncertainty in the limb darkening affects the results only a few per cent. We will compute the factor for a uniform disk where we take the disk radius as a unit.

$$D = \frac{1}{\pi}\int_0^r D_r \cos\gamma \,.\, 2\pi r\, dr = 2D_r \int_0^{\pi/2} \cos\gamma \sin\gamma \cos\gamma \, d\gamma$$

$$= 2D_r \int_0^1 \cos^2\gamma \, d\cos\gamma = \frac{2}{3} D_r \cos^3\gamma \Big|_0^1 = \frac{2}{3} D_r, \qquad p = \frac{3}{2}$$

We define: l = intensity, I = surface brightness (intensity per unit area), σ = surface brightness in magnitude. If σ is a single valued function of the color C, we have $\sigma = f(C)$. If we take two phases where the colors are the same, we see that the light curves show different magnitudes.

$$\Delta m = \Delta m_b = \Delta m_y = \Delta \frac{m_b + m_y}{2} \tag{223}$$

This difference in magnitude is a consequence of the difference in size of the star at these phases. We will now find the connection with the displacement. We have:

$$l_1 = \pi r_1^2 I, \qquad l_2 = \pi r_2^2 I, \qquad \frac{l_1}{l_2} = \left(\frac{r_1}{r_2}\right)^2$$

$$\Delta m' = -5(\log r_1 - \log r_2) = -5\,\Delta \log r = -5 \log e . \Delta \ln r$$

$$= -5 \log e \frac{\Delta r}{r} = -5 \log e \frac{p}{r} \Delta D = -2.17 \frac{p}{r} \Delta D \tag{224}$$

Since D is known at the same phases we can plot $\Delta m'$ against ΔD. Repeating this procedure for other pairs of phases where

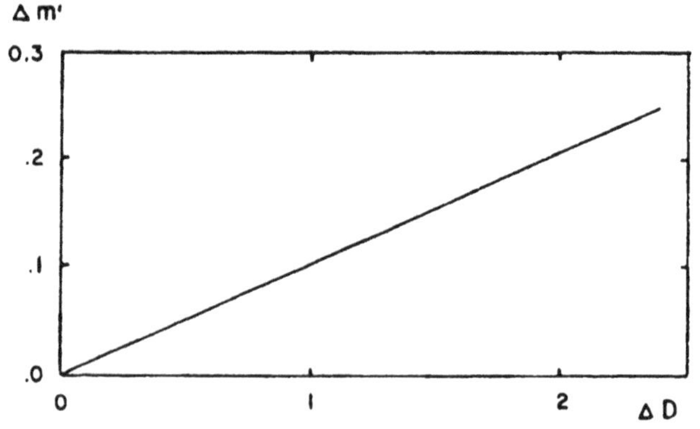

Figure 96. The relation between the magnitude difference at two phases with the same color and the displacement caused by the pulsation of the star.

the colors are the same we find a linear relation. We can derive $p/\bar{r}$ from a determination of the slope, so that $\bar{r}$ follows if a plausible value of p is inserted (Figure 96).

We can now plot the $\Delta m'$ against phase and find the light curve, called here m', which is due to difference of the changing radii only. This curve is proportional to the displacement curve D. Let us now consider the observed light curve in the yellow. For any phase, we have to consider the influence caused by the changing radius or area, and by changing surface brightness thus temperature. We have then in general:

$$l_y = l' I, \qquad m_y = m' + \sigma$$

Because m' is known the σ follows. We can plot σ against phase and find the σ curve, which is due to the difference of changing temperatures only. If we plot σ against the color C it follows from the observations that the relation is linear (Figure 97).

$$\sigma = aC + b = a(m_b - m_y) + b$$

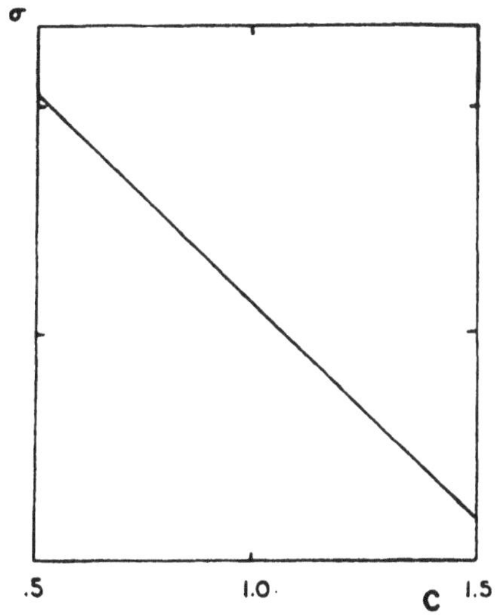

Figure 97. Relation between the surface luminosity in magnitudes and the color of δ Cephei.

The *a* differs from that predicted by the black body theory. We can combine this with:

$$m_y = \sigma - 2.17\frac{p}{r} D + \text{constant}$$

Mere substitution gives:

$$m_y = aC - 2.17\frac{p}{r} D + \text{constant}$$

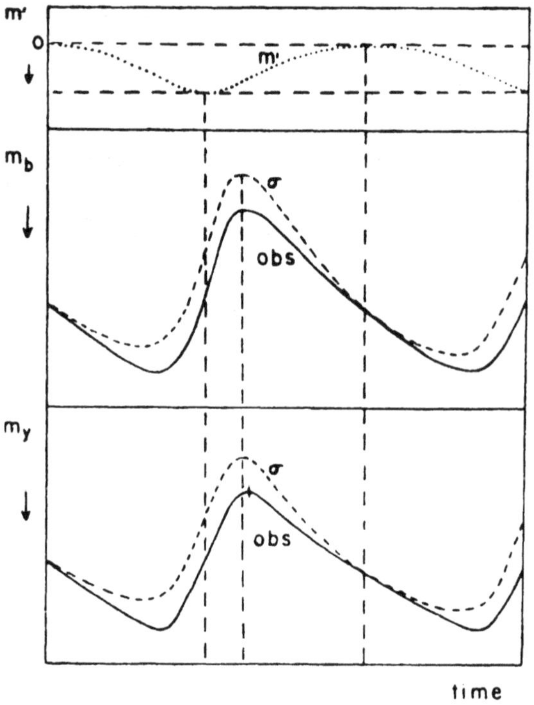

Figure 98. The observed light variation of a Cepheid is caused by the light variation due to changing radius (m'), and by the light variation due to changing surface brightness (σ), thus temperature. The unit chosen here is the maximum area. The observed maximum light happens later at longer wavelengths.

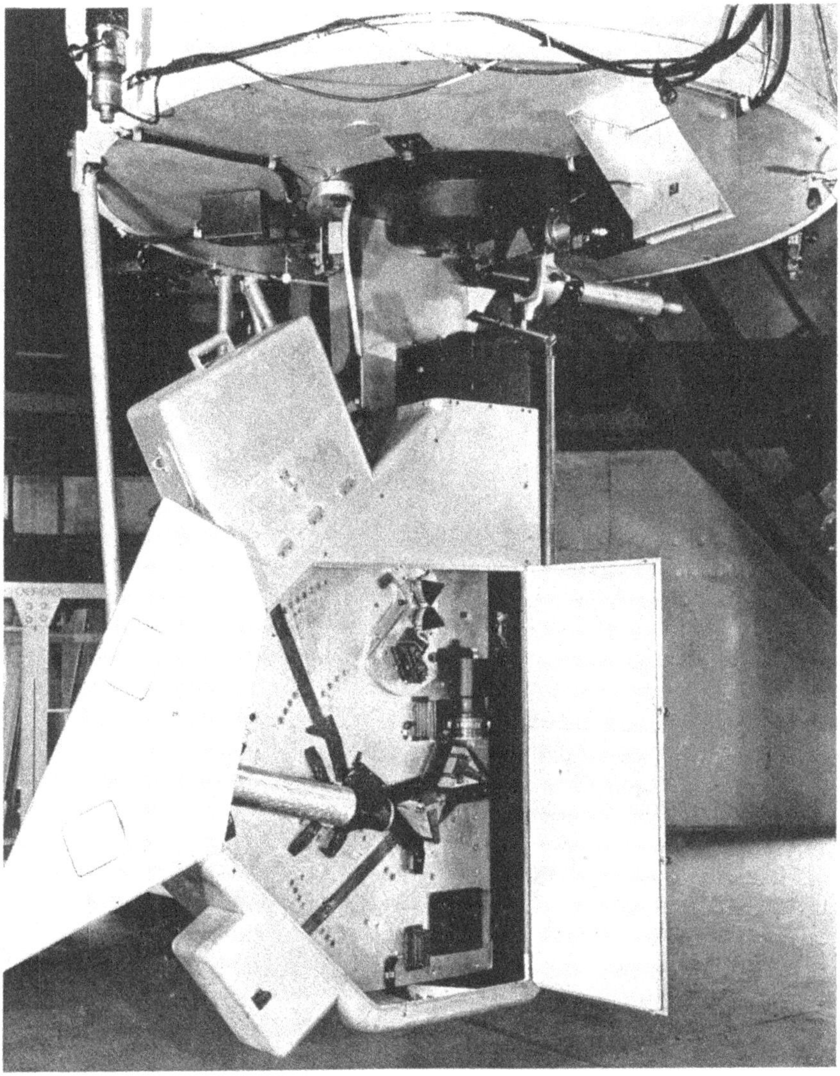

Plate VI. The new spectrograph attached to the 72-inch reflecting telescope of the Dominion Astrophysical Observatory, Victoria. The case has been opened to show the optics inside.
(Courtesy Dr. R. M. Petrie.)

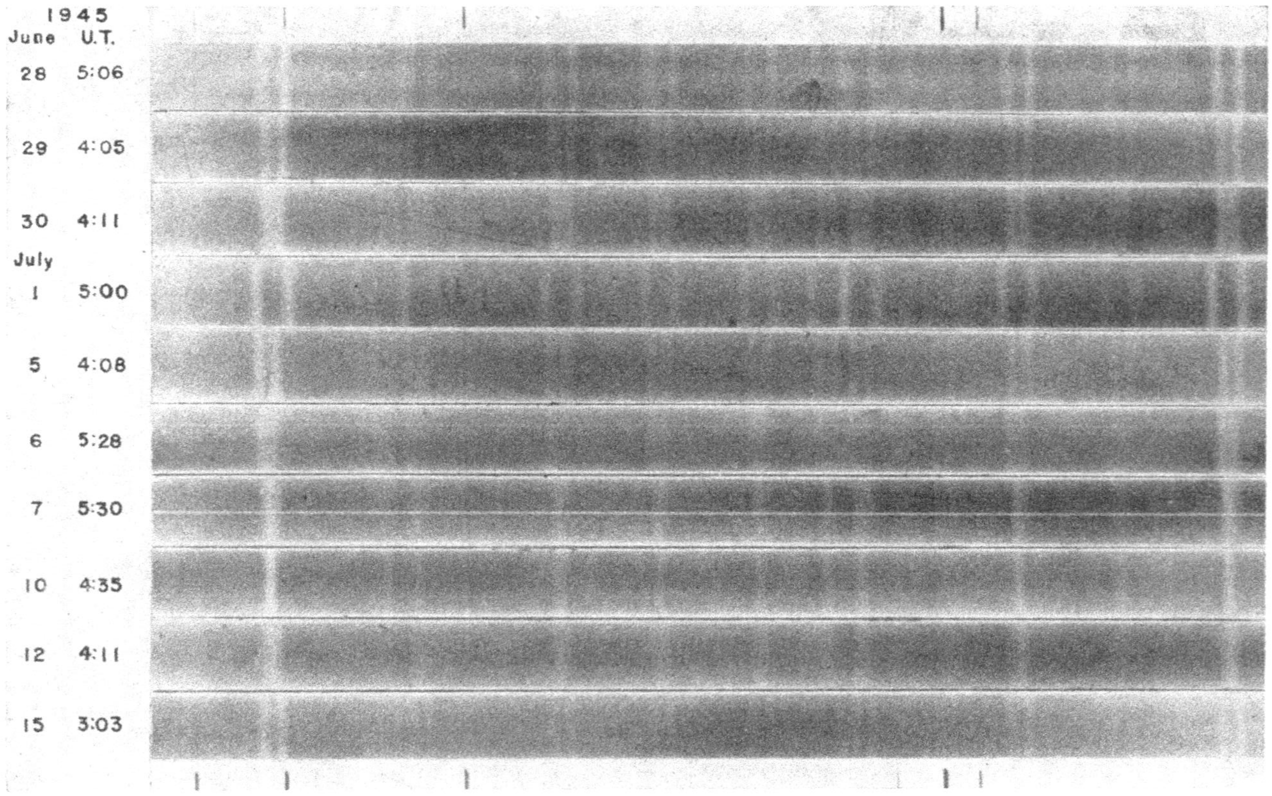

Plate VII. Negative spectra of the spectroscopic double Mizar on ten different dates, showing double lines in the elongations and single lines in the conjunctions. (Courtesy Dr. O. Struve.)

The m_y, C, D are known while the value of a was found from the graph. The unknown quantity $p/\bar{r}$ follows; from this the r is determined. In a least squares solution a can also be taken as unknown, but the equation has to be rewritten somewhat. For δ Cephei one finds, if one takes $p = 1.4$ obtained from the reasonable limb darkening $x = 2/3$:

$$\bar{r} = 37 \times 10^6 \text{ km} = 52\, r_{\odot}$$

The mean radius is thus 52 times the solar radius, thus the Cepheid is a giant. The variation in radius is about 6% in each direction around the mean value, and thus about 12% in total.

The light curves in different wavelengths for the same Cepheid do not have the same times of maxima; the longer wavelengths show a lag. If the surface brightness curves and the area curve are drawn in, it is clear that the former have the same times of maxima, and that the greatest lag occurs for that light curve for which the σ curve has the smallest amplitude. This last situation happens for the curve in the yellow as illustrated in Figure 98.

69. *Period-luminosity relations.* For the Cepheids in the Small Magellanic Cloud a close relation was found between the apparent brightness and the period, or between apparent magnitude and the logarithm of the period. The distance to this cloud was unknown, and the problem was how to find the absolute scale. Therefore for about twenty bright nearby Cepheids the measured proper motions were used to obtain the mean distance of the group. These are thus statistical parallaxes. The Cepheids are giants, and even those which appear brightest are very far away. The zero point determination of the magnitudes was thus quite uncertain.

Globular clusters give a diagram M versus log P in which the cluster variables are plotted. Their absolute magnitude is $M = 0$. Cluster type variables in our neighborhood are called

RR Lyrae variables, and their statistical parallaxes were used for the determination of the zero point. In the diagram the Cepheids in the Magellanic Cloud and in the globular clusters (for example ω Centauri) showed the same slope, so that it was very tempting to consider the relations the same. Those cluster variables and Cepheids are typical standard stars of known candle power. The period and the mean magnitude m being observed gives the absolute magnitude M according to the diagram. Now the parallax follows from:

$$M = m_0 + 5 + 5 \log \pi_y = m_0 + 5 - 5 \log r$$

The m_0 means apparent magnitude corrected for absorption. By observing pulsating stars in extra-galactic systems one could determine the distance of those nearby galaxies. In the more distant ones the systems cannot be resolved into stars.

Recently it was found that this picture is too simple. There are two populations of stars. Population I includes stars in the sun's neighborhood and thus in the spiral arms, and Population II are the stars in the nucleus region of the Milky Way and in the globular clusters. The pulsating stars also show differences. For example for the same period the light curves of members of Population I and Population II are different. The spectral types and the ionized metallic lines, which are a measure of temperature, show that the Population I Cepheids of our neighborhood are more luminous than the Population II Cepheids. There are other arguments concerning the galaxies which make it clear that there are in reality two period-luminosity relations for the Cepheids, one for Population I and another for Population II. For a given period the Cepheids of Population I are about 1.5 magnitudes brighter than those of Population II. The cluster variables and cluster stars are typical Population II stars just like the RR Lyrae variables; their zero point is thus correct because there is only one type of them (Figure 99).

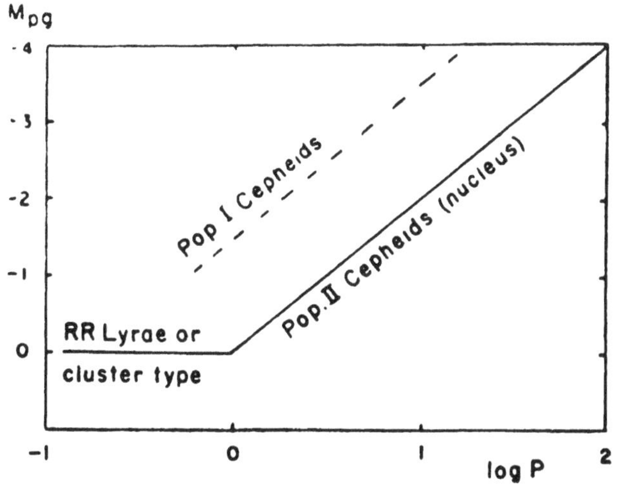

Figure 99. The period-luminosity relations for pulsating stars.

This makes a revision of the old distances necessary. The distances in our Milky Way were determined with help of cluster variables and the size of our Milky Way system was thus determined correctly. The distance of the nearby galaxies were determined via the "classical" Cepheids in their spiral arms, thus by Population I stars. Since they are about 1.5 magnitudes brighter than we had believed, they are about twice as far away as previously considered. This correction increases the scale of the universe. The masses of the extra-galactic nebulae increase by the same factor of two.

REFERENCES

METHOD OF R. LEHMANN-FILHÉS

R. Lehmann-Filhés: *A.N.*, **136**, 17, 1894.
A. A. Rambaut: *M.N.*, **51**, 316, 1891.
F. Schlesinger: *Allegheny Obs. Publ.*, **1**, 33, 1908.

METHOD OF K. SCHWARZSCHILD AND W. ZURHELLEN

K. Schwarzschild: *A.N.*, **152**, 65, 1900.
W. Zurhellen: *A.N.*, **175**, 245, 1907.
R. H. Curtiss: *Publ. Michigan*, **2**, 178, 1916.

METHOD OF J. WILSING AND H. N. RUSSELL

J. Wilsing: *A.N.*, **134**, 90, 1893.
H. N. Russell: *Ap.J.*, **15**, 252, 1902.
W. Zurhellen: *A.N.*, **177**, 321, 1908.
H. C. Plummer: *Ap.J.*, **28**, 212, 1908.
W. J. Luyten: *Ap.J.*, **84**, 85, 1936.

METHOD OF H. N. RUSSELL

H. N. Russell: *Ap.J.*, **40**, 282, 1914.

METHOD OF K. LAVES AND W. F. KING

K. Laves: *Ap.J.*, **26**, 164, 1907.
W. F. King: *Ap.J.*, **27**, 125, 1908.
A. Pogo: *Ap.J.*, **67**, 262, 1928.
J. K. E. Halm: *M.N.*, **87**, 628, 1927.

GAS STREAMS

O. Struve: *M.N.*, **109**, 487, 1949.
O. Struve: *Stellar evolution*. Princeton Un. Press, Princeton, 1950.
O. Struve: *Occasional Notes R.A.S.*, **3**, 161, 1957; *Sky and Telescope*, **17**, 70, 1957.

REFLECTION EFFECT

G. P. Kuiper: *Ap.J.*, **88**, 472, 1938.

RATIO OF INTENSITIES

O. Struve: *Ap.J.*, **72**, 1, 1930.
G. A. Shajn: *Poulkovo Circ.*, No. 22, 1, 1937.
R. M. Petrie: *Publ. Dominion Astroph. Obs. Victoria*, **7**, 205, 1939; **8**, 319, 1950.

CATALOGUES OF ORBITAL ELEMENTS

R. G. Aitken: *The binary stars.* McGraw-Hill Book Co., New York and London, 288, 1935.
J. H. Moore: *L.O.B.*, **18**, 1, 1936.
J. H. Moore and F. J. Neubauer: *L.O.B.*, **20**, 1, 1948.

DIAMETER OF CEPHEIDS

W. Baade: *Mitt. Hamburg*, **6**, 85, 1926.
K. F. Bottlinger: *A.N.*, **232**, 3, 1928.
W. Becker: *Z. f. Aph.*, **19**, 289, 1940; *Mitt. Potsdam No.* 4.
A. van Hoof: *Koninkl. Vlaamsche Acad.*, **5**, No. 12, 1943.
A. J. Wesselink: *B.A.N.*, **10**, 91, 1946; **10**, 256, 1947.
J. Stebbins: *P.A.S.P.*, **65**, 118, 1953.

PERIOD-LUMINOSITY RELATIONS

O. Struve: *Sky and Telescope*, **12**, 203, and 238, 1953.

V

Photometry

THE MOST FUNDAMENTAL RULE IN PHOTOMETRY IS THAT WE should compare two light sources of approximately equal brightnesses and colors. This avoids many troubles which would otherwise be encountered. We shall discuss the visual photometers, and the photographic and photo-electric methods, which give the accuracy required for modern research. For the determination of orbital elements of eclipsing stars accurate light curves are needed.

70. *Intensity.* In physics the practical unit of intensity is the international candle. This unit of intensity is maintained in the various countries by means of such devices as the Hefner lamp, which burns whale oil at a prescribed rate, or by electrically operated carbon filament lamps, or more recently by means of a platinum filled crucible maintained at a temperature equal to the melting point of platinum. Two such candles placed together have an intensity equal to two units. We have to make the distinction between apparent and absolute intensities. The former is defined as the amount of radiant energy that flows per unit time across a unit area placed at right angles to the beam. It depends, therefore, on the distance between the source and the plane of unit area. The absolute intensity of the light source is the radiant energy which it delivers through a given solid angle per time unit. The unit of absolute intensity is defined per

unit solid angle. To compare the absolute intensities of two light sources we can place them at the same distance from us, and measure the ratio of the intensities across unit areas at right angles to the beam. Therefore, we need not be at the source.

Synonyms for intensity are luminosity and brightness. If the radius or the aperture of the objective of a telescope is increased by a factor of 3, the area gain and therefore the apparent intensity gain amounts to a factor of 9.

The inverse square law of intensity, $l \propto 1/r^2$, is to be interpreted as follows. After one second the light energy passes the surface of a sphere of radius c = light velocity; after 3 seconds the radius is $3c$, therefore the surface of the sphere is 9 times as great. Thus on this sphere 9 times less energy passes per unit area than through a unit area on the first one. This holds for all kinds of energy radiating from a point source in a radial direction in a transparent space. This is all quite independent of the units used, and we will therefore postpone the discussion of the astronomical unit of intensity until later. For orbital determination it is clear that in our computations we must use the intensities of the stars involved.

71. *Magnitude*. G. Fechner discovered the general law of sensation, namely that the differences in intensities which correspond to the same fractional part of the whole are equally perceptible to the senses, whether the whole intensity be great or small. Mathematically this law says: Sensation $\propto$ log stimulus. For our case we have $m \propto \log l$ where m is the magnitude and l the luminosity.

Ptolemy called the magnitude of a bright star $m = 1$ and that of a faint star which could just be observed by the naked eye $m = 6$ and was able to estimate the intermediate stars to a quarter of a magnitude. J. Herschel found that this magnitude interval of $\Delta m = 5$ corresponds to an intensity ratio of about one hundred. N. Pogson proposed to define the ratio to be exactly 100 and this Pogson scale is now generally adopted. It determines the relation

$$m = -2.5 \log l \qquad (225)$$

The negative sign indicates that magnitudes increase as the stars become fainter and the intensities smaller. Again we have to distinguish between apparent and absolute magnitudes.

The scale of magnitudes is defined by this equation. The zero point of the visual or photovisual magnitude system is defined to make stars just visible to the naked eye $m = 6$. Because the sensitivity of the eye varies from one observer to another, by international agreement the Polar Sequence of F. H. Seares is followed in this respect. This definition defines also the unit of intensity $l = 1$ or $m = 0$ in astronomy.

We can distinguish magnitudes according to the wavelength in which they are taken. Thus m_{pg} is the blue magnitude for $\lambda = 4430$ A and m_{pv} the yellow magnitude for $\lambda = 5500$ A. Violet and red magnitudes are also used. Both the eye and photographic plate record magnitudes, while the photocell records intensities. The eye is most sensitive in the yellow region; the normal photographic plate is most sensitive in the blue. The magnitude scale is defined by the equation for any effective wavelength. The zero point is defined in such a way that the magnitudes of the blue main sequence stars with spectral type *A*0V coincide with the photovisual magnitudes.

The absolute magnitude, $M = -2.5 \log L$, is defined as the magnitude of the star when seen from a distance of 10 parsecs. For the sun, $M_v = +4.7$. The inverse square law can now be written in logarithmic form.

$$\frac{L}{l} = \frac{r^2}{100}, \qquad M - m = -5 \log r + 5$$

$$M = m + 5 - 5 \log r = m + 5 + 5 \log \pi_y \tag{226}$$

We call $m - M = 5 \log r - 5$ the distance modulus.

72. *Color and reddening.* The color or color index is defined in terms of magnitudes as follows:

$$C = m_{pg} - m_{pv} \tag{227}$$

We have already seen that by definition $C = 0$ for stars of spectral type $A0$V. This is a practical zero point and not derived from the theory of black body radiation. Using wavelengths other than international ones, different color equivalents are obtained. They are usually proportional to the international ones as long as ultraviolet light is excluded, and stars of the same luminosity type are compared.

If a dust cloud could suddenly be placed in front of a star, the spectral type would not change, because the absorption lines of the spectrum originate in the reversing layer of the star. However, we know that the sun appears fainter and redder near the horizon when sunlight must travel through more atmosphere. This shows that the blue sun rays are absorbed and scattered more than the red rays. In a somewhat similar way a star appears fainter and redder when its light passes through an interstellar dust cloud. The color-excess or reddening is defined as the observed color minus the normal color for a star of its spectral type; thus the reddening is also expressed in magnitudes.

$$E = C_{obs} - C_{normal} \tag{228}$$

73. *Black body energy distribution.* Suppose we take two hollow spheres, one of them blackened inside and out by lamp black. Inside each sphere we place some energy source which can be heated by electricity. At low temperature the black sphere will be perfectly black and will be a perfect absorber. The other

sphere will show some color for example grey. At high temperature the black sphere will become hotter and shine brighter than the other sphere; so the black body is also a perfect emitter. This black body radiator is the simplest one in nature, and the only one for which the exact law of radiation can be stated. It is fortunate that in first approximation, stars behave like black bodies. The law was derived theoretically by M. Planck with help of the theory of light quanta. Expressed mathematically we have for the absolute intensity of the radiation per unit area according to the *law of Planck*:

$$I(\lambda, T) = \frac{c_1}{\lambda^5 \left(e^{c_2/\lambda T} - 1\right)} \tag{229}$$

$$c_1 = 1.176 \times 10^{-5}\ \text{erg.cm}^2/\text{sec}, \quad c_2 = 1.438\ \text{cm. deg}$$

This is a relation between three variables, the absolute energy I in ergs, the wavelength λ in cm, and the absolute temperature T in degrees Kelvin. Usually we keep T constant and for the given absolute temperature get the Planck curve, or the relation between intensity and wavelength.

To find out further uses for this curve let us write:

$$x = \lambda T, \qquad \frac{dx}{d\lambda} = T, \qquad d\lambda = \frac{dx}{T}$$

The Planck formula becomes:

$$I(x) = \frac{c_1 T^5}{x^5 \left(e^{c_2/x} - 1\right)} \tag{230}$$

The entire area under the curve is then:

$$I_{\text{bol}} = \int_0^\infty I(\lambda, T)d\lambda = \int_0^\infty I(x)\frac{dx}{T} = \int_0^\infty \frac{c_1 T^5}{x^5 \left(e^{c_2/x} - 1\right)} \frac{dx}{T}$$

$$= T^4 \int_0^\infty \frac{c_1 dx}{x^5 \left(e^{c_2/x} - 1\right)} = \sigma T^4$$

The computation of the integral determines the constant. By integrating the Planck formula we find this *law of Stefan and Boltzmann* which states that for a black body the total or bolometric energy generated in all wavelengths together is proportional to the fourth power of the absolute temperature.

$$I_{\text{bol}} = \sigma T^4, \qquad \sigma = 5.672 \times 10^{-5} \text{ erg/cm}^2\text{sec.deg}^4 \tag{231}$$

To measure the Planck curve and the total energy we need a radiation detector which has equal sensitivity in all wave lengths. There are only two such detectors commonly used in astronomy, the bolometer and the thermocouple. In the bolometer the electrical resistance changes when radiation is incident. The thermocouple consists essentially of two different metals joined together. If energy of any wavelength hits this joint, a voltage difference occurs which can be measured by electrical means.

For small λT the exponent of e becomes large in the Planck formula so that we can neglect the one in the denominator without making a serious error. This gives the Wien curve, which consequently is only an approximation but historically it was found earlier than the Planck curve.

$$I(\lambda, T) = \frac{c_1}{\lambda^5 \, e^{c_2/\lambda T}} \tag{232}$$

Figure 100 shows that both curves coincide in the visual region but that the Wien curve is somewhat lower in the infrared. Observation in that region shows that the Planck curve is the correct one. We would like to find the maximum of the Wien curve.

$$\frac{dI}{d\lambda} = c_1 e^{-c_2/\lambda T}\left(- 5\lambda^{-6} + c_2 \frac{\lambda^{-7}}{T}\right)$$

For the maximum this must be zero.

$$\frac{dI}{d\lambda} = 0 = -5\,\lambda^{-6} + c_2\,\frac{\lambda^{-7}}{T}, \qquad 5\lambda = \frac{c_2}{T}$$

$$\lambda_{\max} = \frac{c_2}{5\,T} = \frac{0.2876}{T}$$

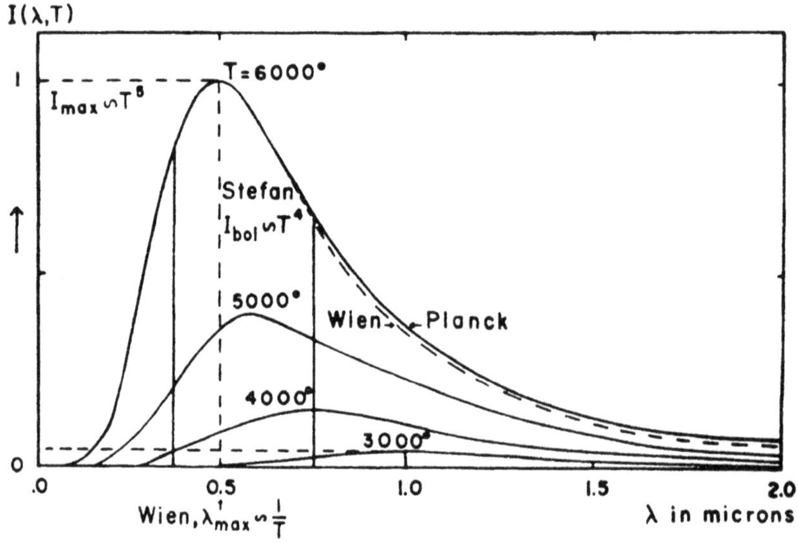

Figure 100. Planck curves for various absolute temperatures. The coordinates of the maximum are respectively $\lambda_{\max} \propto 1/T$ (Wien's law) and $I_{\max} \propto T^5$. The entire area under the curve or the bolometric intensity is $I_{\text{bol}} \propto T^4$ (Stefan's law).

A more accurate value is found by differentiating the Planck curve in a similar way. For the *Wien displacement law* we then find:

$$\lambda_{\max} = \frac{A}{T}, \qquad A = 0.2897\ \text{cm. deg} \tag{233}$$

The wavelength of maximum intensity is inversely proportional to the absolute temperature. The ordinate corresponding to the maximum is found by substitution.

$$I_{\max} = \frac{c_1 T^5}{A^5 e^5} \propto T^5 \tag{234}$$

The maximum intensity is proportional to the fifth power of the absolute temperature. Thus as the absolute temperature becomes twice as great, the wavelength of maximum intensity is halved and shifted to the blue; the maximum intensity increases 32 times; the total intensity over all wavelengths becomes 16 times as large as before.

Stars approximate black body radiators very closely. If there are small deviations, the stars are termed grey body radiators. This offers us three ways in which to determine the temperature of a star, namely solving for T in each of the laws of Planck, Stefan, and Wien. Naturally here we have to consider the continuum of the spectrum and not the absorption lines.

As λT becomes large, we get the Rayleigh-Jeans formula:

$$I(\lambda, T) = \frac{c_1}{\lambda^5} \frac{1}{(1 + c_2/\lambda T - 1)} = \frac{c_1}{\lambda^5} \frac{\lambda T}{c_2} = \frac{c_1}{c_2} \frac{T}{\lambda^4} \tag{235}$$

Thus for large λT the luminosity is proportional to the absolute temperature and inversely proportional to the fourth power of the wavelength. In the visual region the Rayleigh-Jeans curve deviates considerably from the Planck curve; however, in the radio region it is a good approximation.

If we convert intensities into magnitudes, we find the following relations (Figure 101):

$$m = 12.5 \log \lambda + 2.5 \log (e^{c_2/\lambda T} - 1) + \text{constant}, \quad \text{Planck}$$

$$m = 12.5 \log \lambda + 2.5 \frac{c_2}{T} \log e + \text{constant}, \quad \text{Wien}$$

$$m = 10.0 \log \lambda - 2.5 \log T + \text{constant}, \quad \text{Rayleigh}$$

If we increase the absolute temperature by a factor of 2, the

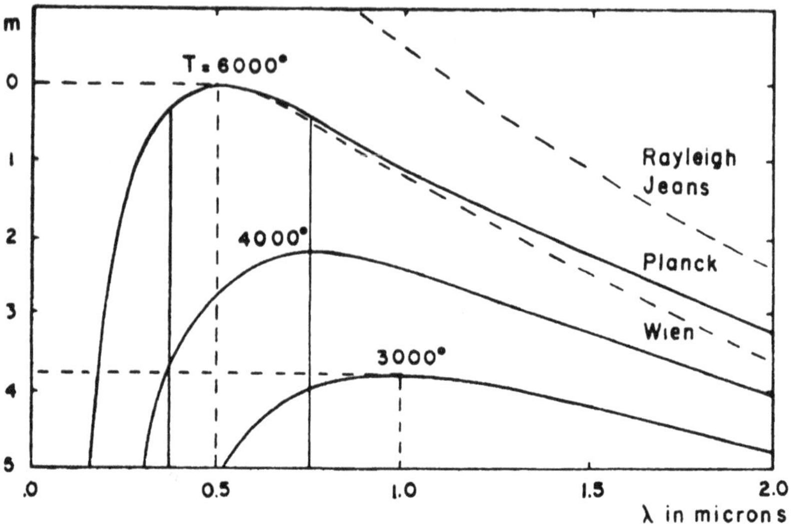

Figure 101. The Wien, Planck and Rayleigh-Jeans curves for $T = 6000°$ together with two other Planck curves for $T = 4000°$ and $T = 3000°$.

wavelength of maximum light becomes smaller by a factor of $\frac{1}{2}$, while the maximum magnitude becomes $-2.5 \log 32 = -3^m\!.75$ brighter. In the visual region the curvature of the Planck curve is slight as seen in Figure 102.

The radiation curves for different temperatures do not intersect except upon the λ axis for $\lambda = 0$ and $\lambda = \infty$. Thus for a given I (or m) and λ which correspond to one point in the diagram we find only one curve of a specified T. In other words the T follows and therefore the color. This holds for the case where I or m is measured absolutely.

In astronomy we could do the same working with absolute intensity or absolute magnitude. However, one needs to introduce the distance of the star to compute this, and the inaccuracy

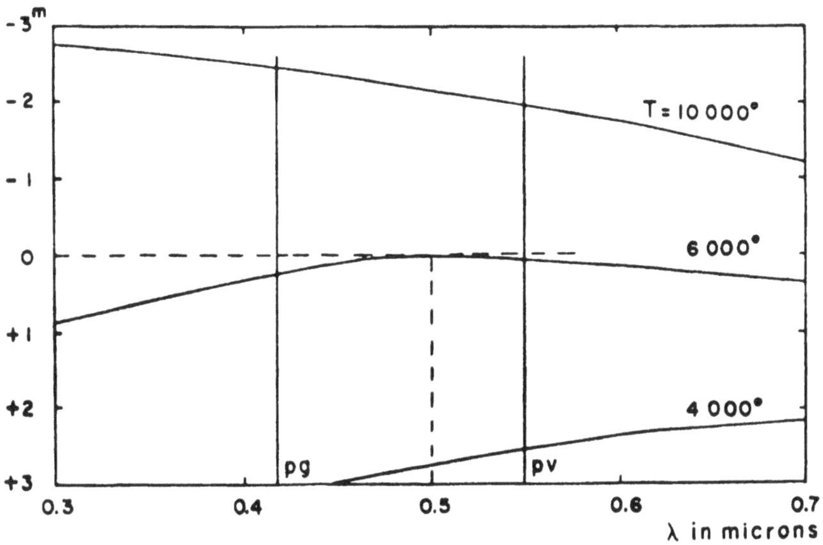

Figure 102. Three Planck curves in the visual region.

in this distance would affect the result. A changing distance means a shift of the whole curve in the magnitude direction. Therefore, in practice, instead of this distance we introduce a second I (or m) at another wavelength. Those two points in the diagram define the curve and thus the absolute temperature and the normal color for the case of apparent intensity or magnitude. If these two points are taken very close together we get in the limiting case the gradient, which is thus also a color equivalent. If there is interstellar reddening, the form of the curve changes and thus alters the observed color and the gradient. One needs another datum, namely the spectral type, to define the normal color or the normal gradient for the corresponding temperature. Now we can determine the amount of reddening. One needs a third intensity or magnitude at another wavelength to determine whether there are deviations from the black body radiation in the real starlight.

It is sometimes convenient to give m as a function of $1/\lambda$. The Wien curve then becomes:

$$m = -12.5 \log \frac{1}{\lambda} + \frac{1.558}{T} \frac{1}{\lambda} + \text{constant} \tag{236}$$

The gradient $\dfrac{dm}{d(1/\lambda)}$ is proportional to $\dfrac{1}{T}$ and is therefore a temperature equivalent. The color index for the same star becomes:

$$C = m_{pg} - m_{pv} = 12.5 \log \frac{\lambda_{pg}}{\lambda_{pv}} + \frac{1.558}{T} \left(\frac{1}{\lambda_{pg}} - \frac{1}{\lambda_{pv}} \right)$$

$$C = a + \frac{b}{T} \tag{237}$$

The color varies linearly with $1/T$ so that a large color index corresponds to a cool star. As we have seen, the spectral type is associated with $\log T$. Though both color and spectral type are temperature equivalents or indicators, the relation between them is not linear.

The absorption and reddening of star light are caused by interstellar dust particles which in the first approximation work according to a $1/\lambda$ law in interstellar space. In the far ultraviolet and in the infrared there is a deviation from this simple relation as shown by observations of J. Stebbins and A. E. Whitford. In the visual region this is thus different from the $1/\lambda^4$ Rayleigh law for the scattering of sun light and star light in the terrestrial atmosphere. In the latter case of scattering, molecules are the cause; in the former, dust particles or grains of a size equal to the wavelength of visible light are responsible. In the plot m against $1/\lambda$ we can easily see the influence of a dust cloud on the distribution of star light in the spectrum (see Figure 103). In the upper curve the normal color is $C_{\text{normal}} = m_{pg} - m_{pv}$. In the

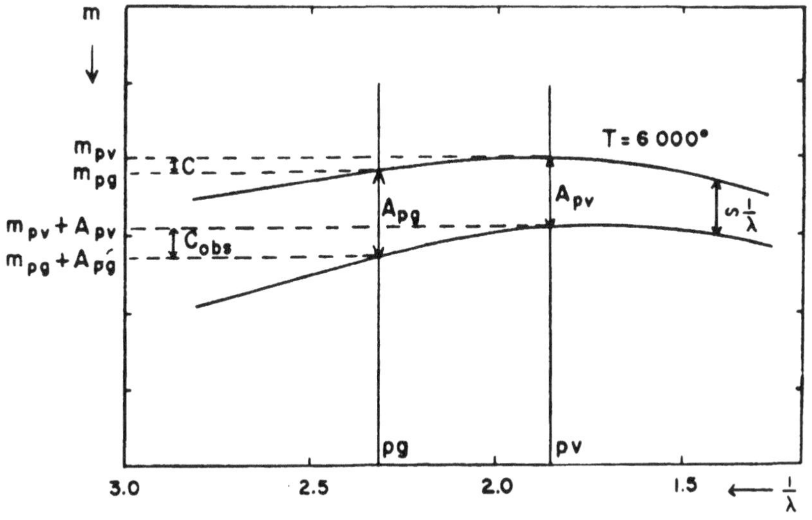

Figure 103. The upper curve is the Planck curve for $T = 6000°$. The lower curve is after passage of the light through a dust cloud.

lower curve the star is fainter and more so in the blue than in the red because the absorption depends on the wavelength. The observed magnitudes are $(m_{pg} + A_{pg})$ and $(m_{pv} + A_{pv})$ respectively, and the observed color is therefore:

$$C_{obs} = (m_{pg} + A_{pg}) - (m_{pv} + A_{pv}) = (m_{pg} - m_{pv}) + (A_{pg} - A_{pv})$$
$$C_{obs} = C_{normal} + E$$

The last relation follows from the definition of color excess or reddening. We can write for this reddening E therefore:

$$E = A_{pg} - A_{pv} \tag{238}$$

The reddening is caused by the difference in selective absorption. Further, the ratio A_{pg}/E is an important constant in our Milky Way system, showing that in the largest part of this system the dust particles are of the same average size.

74. *Observed energy distribution.* Let us consider what we observe in practice and what we really measure. We will first consider what happens between the star and the instrument. The star radiates only approximately as a black body, more accurately as a grey body, but even that is only an approximation. In the continuum of stellar spectra there are many absorption lines which affect the curve. Monochromatic magnitudes must therefore be taken between these lines if possible. A more serious difficulty is the so-called Balmer limit in the ultraviolet (3650 A); for shorter wavelengths the Balmer absorption cuts off part of the continuum. Ultraviolet magnitudes may therefore be affected. Between the star and us there may be an interstellar

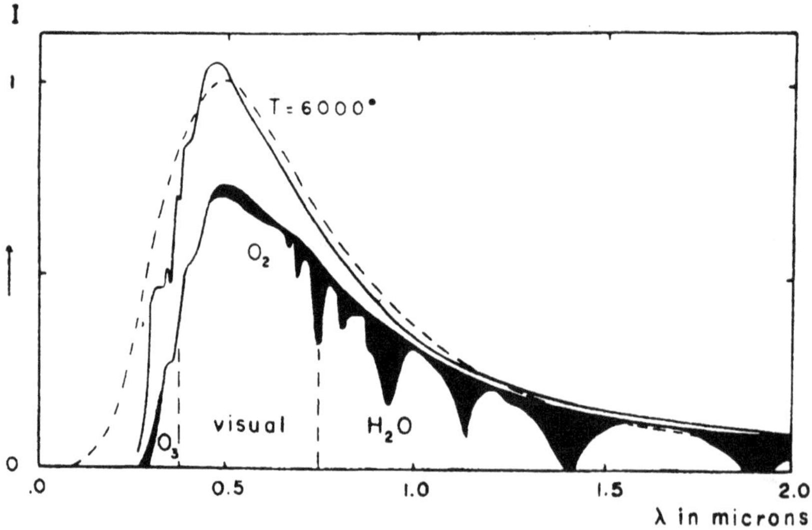

Figure 104. The lower curve represents the radiation of the sun as observed at the surface of the earth showing the absorption caused by water vapor, oxygen and ozone in the atmosphere. The higher curve is reduced to outside atmosphere and compared with the black body radiation at $T = 6000°$.

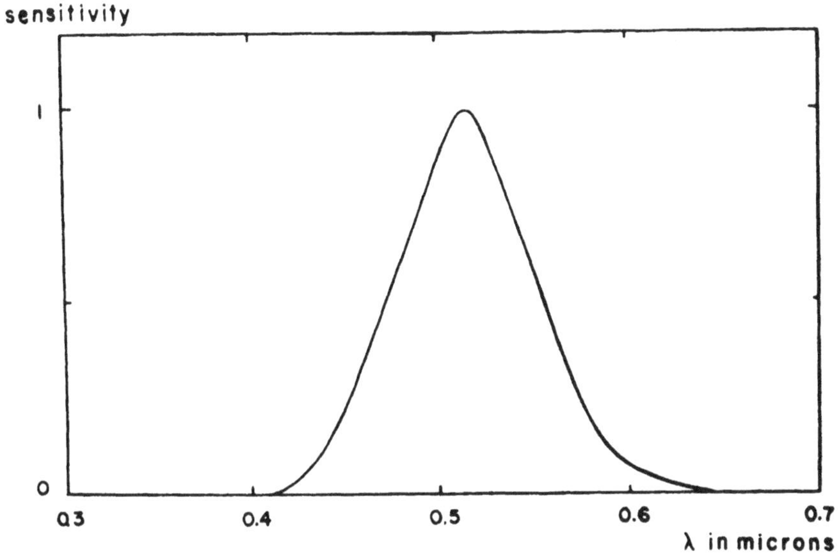

Figure 105. The sensitivity curve for the eye adapted to a faint source of equal intensity in all wavelengths.

dust cloud which affects the curve in the manner already described.

When the star light enters our atmosphere, it is absorbed and reddened again but as we have seen in a somewhat different fashion. As far as the continuum is concerned, these effects are respectively extinction and color extinction. They depend on the air mass traversed and can be determined by observing the star at different zenith distances. Both change from night to night. The far ultraviolet is almost completely absorbed; absorption bands of O_3, O_2, H_2O also appear in the visual region. In the infrared the water vapor bands are extremely deep. For still longer wavelengths they cut off almost all intensity, and only a few windows are left. In the radio region, fortunately, we have a large window (Figure 104).

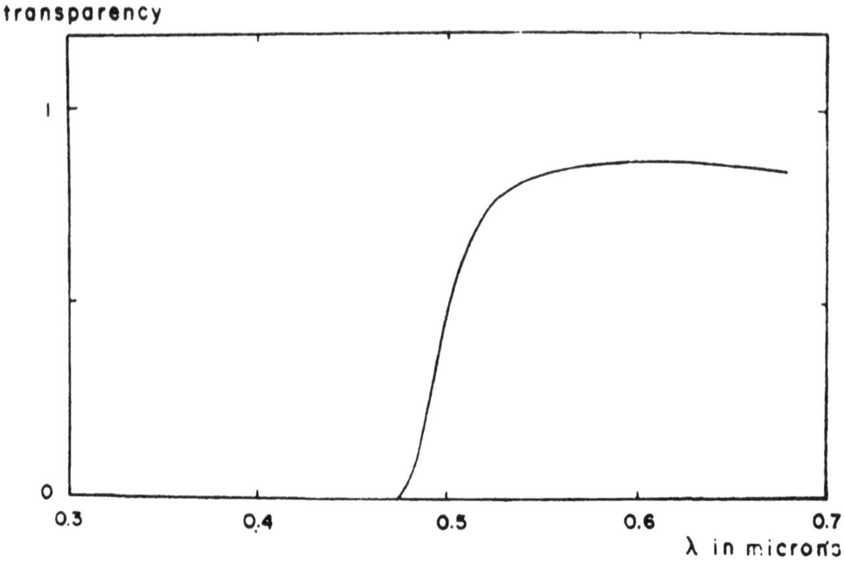

Figure 106. The transparency curve for a yellow filter.

This light now reaches the instrument. The reflection from a mirror is almost the same for all wavelengths, but an average objective absorbs all the ultraviolet light shorter than 3600 A depending on the amount of metal used in the glass. A quartz lens transmits this ultraviolet light. The beam now strikes the measuring instrument. A thermocouple or a bolometer gives a deflection proportional to the intensity at any wavelength, but either is quite insensitive. They are used especially for infrared or heat waves. The eye is most sensitive in the yellow, and we have to consider the so-called sensitivity curve, which gives us the effectiveness or percentage of sensitivity for different wavelengths (Figure 105). It is clear that the incident intensity of a certain wavelength has to be multiplied by a factor corresponding to this wavelength to give the effective result. Then this must

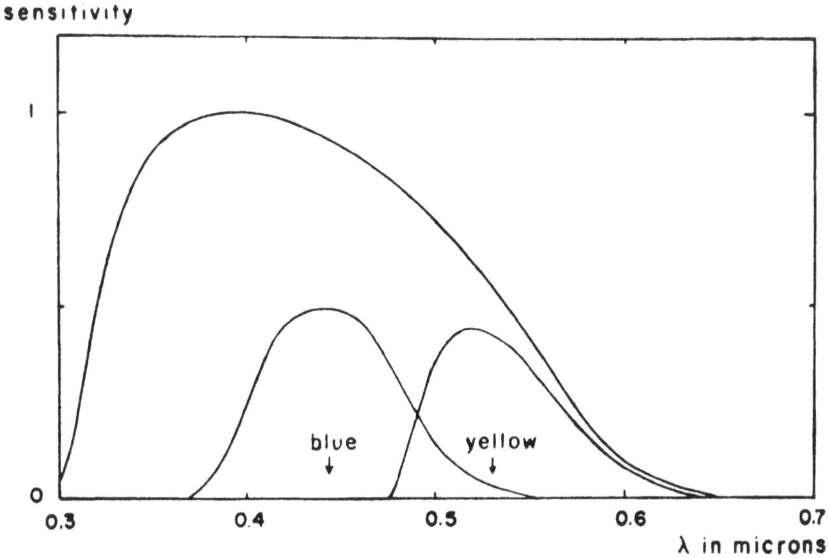

Figure 107. The sensitivity curve for the 1P21 multiplier photocell. The lower two curves are the combination of the 1P21 and a yellow and blue filter respectively.

be converted into magnitudes because magnitudes are what the eye sees. For a photographic plate we must consider the sensitivity curve of the plate and proceed in the same way. A spectrum plate shows magnitudes naturally; calibration of density marks is therefore necessary for photometry on these plates. Also in stellar photometry spectral plates are customarily taken with help of prisms; in this case the linear scale is not proportional to the wavelength.

When the light reaches a photo-electric cell, which measures intensity, we must consider the sensitivity of the cell, which may be most sensitive in the blue region. Again we have to take into account the effectiveness factor for each wavelength. Often color filters are used; each has a transparency curve giving the fraction of light of each wavelength which the filter transmits. We can determine this curve with a quartz monochromator, which is a spectroscope with a second movable slit placed in the plane where the spectrum is formed (Figure 106). The filter-plate or filter-cell combination mainly determines the effective wavelength of the system (Figure 107). This can easily be seen if the incident light in each wavelength has equal intensity. The effective wavelength depends slightly on the temperature and thus color or spectral type of the star, because the gradient of the radiation curve is different for stars of different temperatures. It is also clear that such a combination is not strictly monochromatic, but contains an interval of the spectrum which may include some absorption lines.

75. *Visual method.* Visual estimates are not sufficiently accurate to be used for a precise orbital determination. However, relative observations made with certain types of visual photometers can still be used today, for example in determinations of times of minima and in light curves where the primary minimum is deeper than 0.6 magnitude.

In a polarizing photometer, one Nicol prism and one Iceland crystal are placed in the light path. The light is plane polarized and by optical means the ordinary beam from the variable star is placed alongside the extraordinary beam of the comparison star or vice versa. The intensities of the images are thus varied in opposite directions by rotation of the Nicol prism. There are four positions of equal intensity, separated by an angular amount which depends on the relative intensities of the two stars. To eliminate errors caused by the change in sensitivity across the retina of the eye, the measures are then repeated with the images interchanged in position. Finally, one reverses the entire instrument and repeats the entire process, so that in total one makes 16 measurements of equal intensity. The advantage is that one compares the star images. An experienced observer will complete such a set in about 5 minutes; the average value has a probable error of about ± 0.04 magnitude.

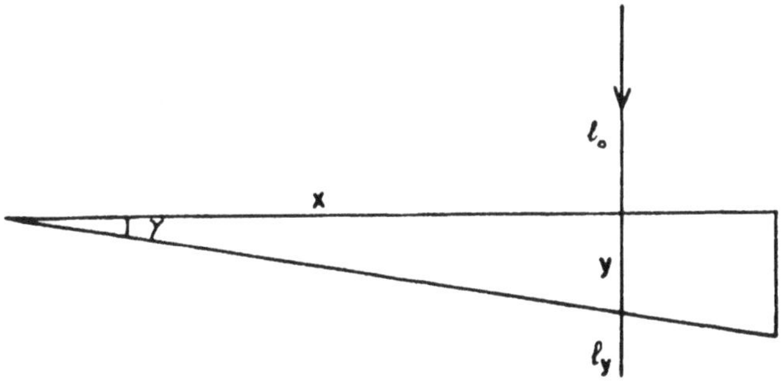

Figure 108. The wedge in the wedge photometer gives an absorption in magnitude proportional to the distance *x*.

The wedge photometer is another useful instrument (see Figure 108). If l is the intensity of a light beam incident on the wedge the absorption or extinction will be $-dl$. The fractional loss of light will be proportional to the thickness dy. We have:

$$-\frac{dl}{l} = a\,dy, \qquad \int_0^y \frac{dl}{l} = \int_0^y -a\,dy$$

$$\ln l_y - \ln l_0 = \ln \frac{l_y}{l_0} = -ay$$

$$\frac{l_y}{l_0} = e^{-ay} = c^y \tag{239}$$

This is a general formula which holds just as well for glass plates with parallel sides or for parallel layers of air. If we call x the distance to the edge and γ the angle of the wedge we have in our case:

$$\frac{l_y}{l_0} = c^{x \tan \gamma}, \qquad \Delta m = -2.5\,x \tan \gamma \log c = K\,x \tag{240}$$

The absorption in magnitudes caused by the absorbing wedge is proportional to the distance between the edge and the incident light. In addition one places a wedge of transparent glass in such a way that both wedges form a plane parallel glass plate and thus the geometrical path does not change appreciably. The proportionality constant is usually calibrated with help of a known magnitude sequence; deviations from the expected linearity will show up at the same time. An artificial light source is used in these observations which is supposed to keep constant light during the observations.

76. *Photographic method.* K. Schwarzschild was one of the first to discuss photographic effects on a negative plate. These plates are always used because positive plates would introduce an additional step and therefore additional plate errors in the process. The density w of a photographic image is a function of the intensity l of the original light source and of the exposure time t according to $w = f(l, t^p)$, where $p = 0.8$ approximately. If two intensities, l_1 and l_2 give the same densities of photographic images in exposure times t_1 and t_2, we have:

$$l_1 t_1^p = l_2 t_2^p, \qquad \log \frac{l_1}{l_2} = p \log \frac{t_2}{t_1}$$

Converting into magnitudes we find:

$$m_1 - m_2 = -2.5 p \log \frac{t_2}{t_1} = 2 \log \frac{t_1}{t_2} \tag{241}$$

For $t_1 = 3t_2$ we get $m_1 - m_2 = 1$. Thus we will reach one magnitude fainter if we expose three times as long. For modern plates p is not always constant, but the rule is still a good approximation.

In earlier times scale determinations were made by increasing the exposure times in a logarithmic scale according to the Schwarzschild's rule, by using the logarithm of the measured diameters, or by rotating disks. These procedures do not give very accurate results. In more recent times neutral filters and gratings have been used to diminish the light of a star by a known amount. A neutral filter giving the same amount of absorption for all wavelengths is still difficult to make; objective gratings also give trouble and must be used with the right combination of telescope and measuring photometer. Steady transparency during the exposure time is a necessity for accurate photometry.

In practice the zero point of the magnitude system is determined by comparison with stars in the North Polar Region. For an orbit computation where we find relative elements, the scale of magnitudes is essential, but the zero point is not. When we observe a variable star, the color of the comparison stars should be the same as that of the variable because of the effect of atmospheric dispersion. The images of the stars are small spectra; water vapor in the atmosphere may absorb some parts of the spectra. Another stipulation of the comparison stars is that their collective range in magnitude must cover the entire light variation of the variable star.

77. *Objective grating*. An objective grating consists of parallel bars mounted in a circular frame. Instead of a single image on the plate we get a central image with first and higher order spectra to both sides. If the plate is taken a little interfocally, these spectral images look just like star images.

Because of the wave structure of light we get interference. If the difference in path length is λ, we will see an image. We have in Figure 109:

$$\lambda_{\text{eff}} : (b+s) = \frac{u}{2} : f, \qquad \lambda_{\text{eff}} = \frac{(b+s)\,u}{2f} \tag{242}$$

Here b = bar width, s = opening, λ = wavelength, f = focal length, u = distance between first order spectra.

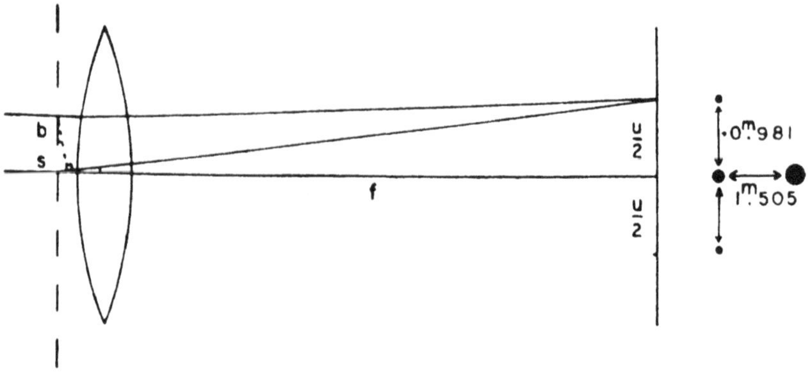

Figure 109. The objective grating number one.

If we use grating one, in which the bars and spaces are of equal width, we get $\lambda_{\text{eff}} = bu/f$. If we measure u on the plate, we can find the effective wavelength which is a measure of the color, but a better color determination is given by the color index. The grating formula which gives the intensities of the images is:

$$\frac{s}{s+b} = \left(\frac{s}{s+b}\right)^2 + 2\sum_{n=1}^{\infty}\left(\frac{\sin n\pi \dfrac{s \text{ or } b}{s+b}}{n\pi}\right)^2 \qquad (243)$$

The left-hand side gives the intensity which passes through the spaces of the grating, while the first term on the right-hand side gives the intensity of the central image, and the remaining terms the intensities of the succeeding spectral images. For grating one, where we take the intensity of an image without a grating equal to unity we have therefore:

$$\frac{1}{2} = \frac{1}{4} + 2\left(\frac{1}{\pi^2} + 0 + \frac{1}{9\pi^2} + \dots\right), \quad \text{for } n = 0, 1, 2, 3, \dots$$

It is easily seen that one-half of the incident intensity passes through the grating, that half of this is concentrated into the central image, and that the second order spectrum does not exist. If we convert into magnitudes, we see that the central image must be $1\overset{m}{.}505$ fainter than the image taken without the grating, and that each of the first order spectra is $0\overset{m}{.}981$ fainter than the central image. The difference is approximately one magnitude, which explains the name "grating one".

The procedure followed in practice is as follows. Take several exposures on one plate alternately with and without the grating. We adopt the theoretical value of $1\overset{m}{.}505$ as the difference between the central image and that taken without a grating. Both are real star images; we thus know the magnitude difference which agrees with the measured density difference on the negative plate. We should find $0\overset{m}{.}981$ as the difference between first order spectra and the central image, but in practice we often find a somewhat smaller value such as $0\overset{m}{.}97$, which we call the grating constant. This may be due to the spectral appearance of the first order images. In the reduction process we always use a magnitude difference of 1 and afterwards multiply by the grating constant factor 0.97. Now we are ready to obtain the scale of magnitudes on the photographic plate.

78. *Rich star field.* In this absolute work we can take plates of an open cluster to determine the magnitudes. This we do both with and without a grating on the same plate. We should not observe more than three hours out of the meridian, because otherwise the relative extinction effect over the plate becomes too great. Photographic extinction is approximately twice visual extinction. For this case this approximation is satisfactory because we apply only relative extinction. Lines of equal relative extinction run parallel to the horizon at the time of exposure.

For the zero point determination, we take on one plate first the star field and then the North Polar Sequence. On one of the two occasions we make a double exposure to help the identification. The star field is taken at a time such that the zenith distance of the field is the same as the constant zenith distance of the North Polar field. This avoids applying a correction for the difference in extinction between the two fields. The magnitudes of the North Polar Sequence have been derived at the Mount Wilson Observatory and reduced to that pole. Our magnitudes will thus be reduced to the Mount Wilson pole, for which the zenith distance is 54°.

79. *Measurement.* We measure a plate with help of a suitable photometer. A light beam passing through the plate falls on a thermo-element or a photocell. This is connected in an electrical circuit, and the deflection of a galvanometer can be read from a scale in centimeters. For no transmitted light at all we get no galvanometer deflection; for light through the plate fog we make the deflection 25 cm, for example, by means of a variable resistance. A very black star image intercepts the light beam, and we get a deflection around 2 cm; very faint images give a deflection of about 23 cm; all measurements are made between these two limits (Figure 110).

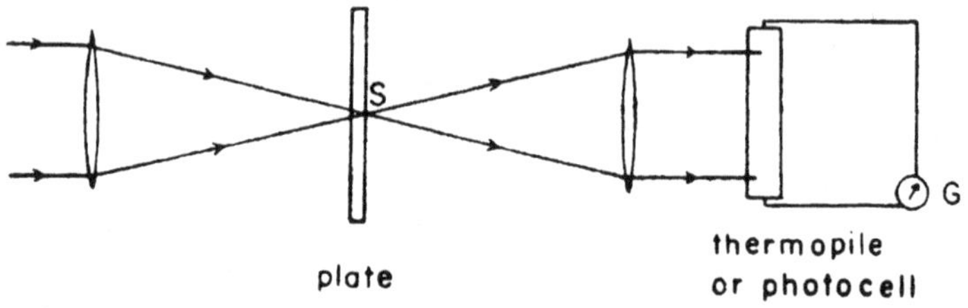

Figure 110. The principle of the Schilt photometer for measuring the magnitudes of photographic images on a negative plate.

There are also other types of useful photometers, such as one which uses a variable iris diaphragm. The diameter of the diaphragm is set and measured so that we get a constant deflection.

If I_0 is the intensity or luminosity of the light beam before, and I_y the intensity of the light beam after it passes through a plate, we can define the following quantities:

$$\frac{I_y}{I_0} = \text{transparency}, \quad \frac{I_0}{I_y} = \text{opacity}, \quad \log \frac{I_0}{I_y} = \text{density}$$

80. *Reduction.* For each star on the plate we plot the deflection of the central image against the mean deflection of the two first order spectra (Figure 111). If we draw the 45° line, we know that

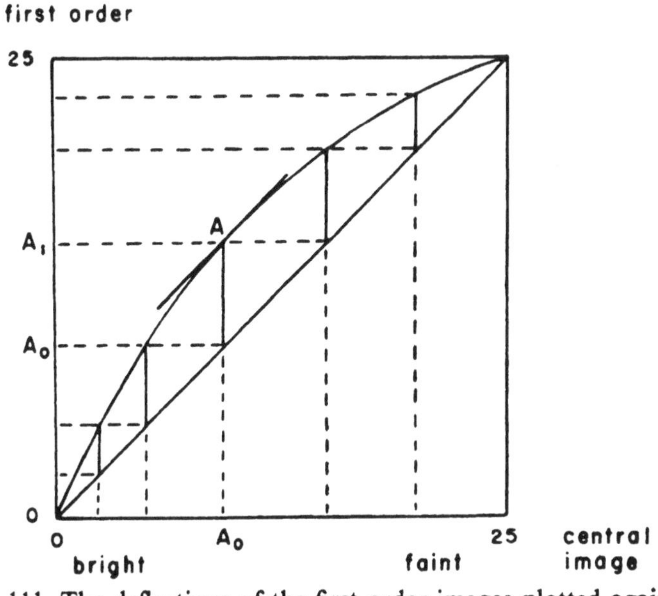

Figure 111. The deflections of the first order images plotted against the deflection of the central image.

the difference in galvanometer deflections corresponds to a difference of one magnitude. We start by writing the coordinates of a point A, and by following the zigzag line we can find other points that have just the same difference of one magnitude. In the figure we thus find six ordinates and correspondingly five magnitude differences.

The so-called reduction curve is now constructed in the following way. From theoretical considerations by K. Schwarzschild, it follows that the middle part of this curve is a straight line (Figure 112). We position it so that the deflection difference $A_1 - A_0$ agrees with $\Delta m = 1$. The other points follow easily.

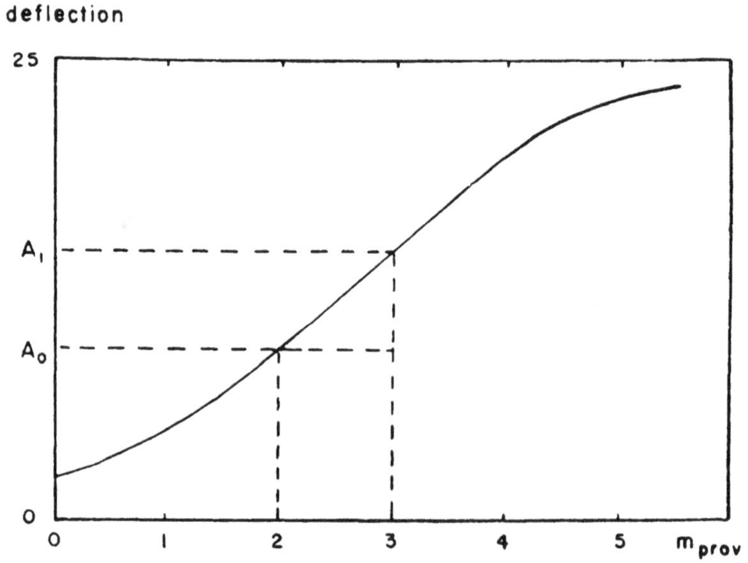

Figure 112. The reduction curve for deriving provisional magnitudes.

By changing the first ordinate we can find more points on the reduction curve in this fashion. We now convert the deflections of each star to the corresponding provisional magnitudes. Multiplying these by the grating constant determines the magnitude scale for the plate. The zero point cannot be found in this way. We must first bring all plates of the region to the same provisional zero point and then connect this to the zero point of the North Polar Sequence.

The slope of the curve is a measure of the gradation of the plate. It can be seen that the gradation is small if a large magnitude interval is measurable. This gradation is constant for one exposure of the different stars on a plate. Due to a change of seeing conditions it may change slightly for a second exposure. A second plate will always have a different gradation by virtue of different developing conditions. From a great number of plates reduced in this manner, the mean reduction curve can be found and given in the form of a table. These table magnitudes will be used later. It is also possible to replace the centimeter scale of the above mentioned galvanometer by a magnitude scale after enough reductions have been completed.

81. *Magnitude systems.* We have arrived at a magnitude system for a certain effective wavelength, and we would like to compare it with the results found for the same field by other observers. We can expect a zero point difference, a scale factor, and a color equation between the two systems. This last is caused by the different instruments, plates, and cells used, each of which may have different color sensitivities. The expected relation is thus of the form (Figure 113):

$$m' = a_1 + b_1 m + c_1 C \tag{244}$$

When m and C are very accurate this can be used in a least squares solution to determine a_1, b_1, c_1. When both magnitude systems have comparable accuracies, the equation of condition is written in the form:

$$m' - m = a_2 + b_2(m' + m) + c_2 C \tag{245}$$

The solution also determines the probable errors. The constant a_1 is close to zero, b_1 approximately one, and c_1 close to zero when the effective wavelengths are not very different. The sign of

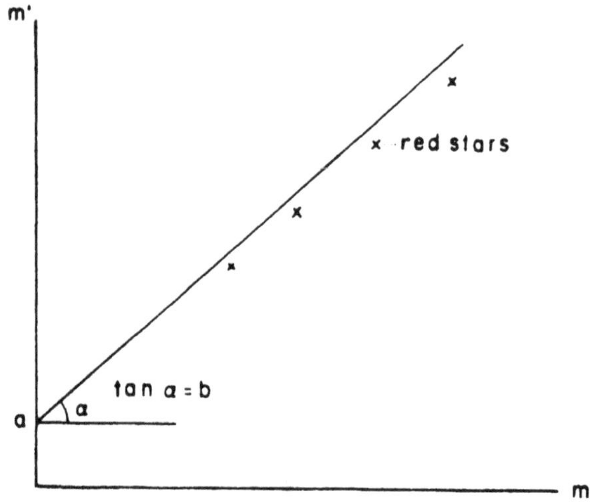

Figure 113. Example of relation between two magnitude sequences.

c_1 tells in which direction the wavelengths differ from each other.

It is clear that one needs stars of a large color range at any small magnitude interval to determine the c_1 accurately. In the case of open clusters which contain no giants, one should observe a few bright red field stars in addition to the cluster stars for such a comparison.

82. *Discovery of variables.* The first variables were found by the naked eye. The variability of Mira Ceti was discovered by Fabricius in 1596. Algol was discovered in 1670 by Montenari, and in 1782 J. Goodricke, a young English deaf mute, first determined the period and suggested the correct cause. It is possible, therefore, to discover a variable visually just by comparing its light with that of one or more comparison stars over a period of time. It is clear that for this the light variation must be of the order of one magnitude.

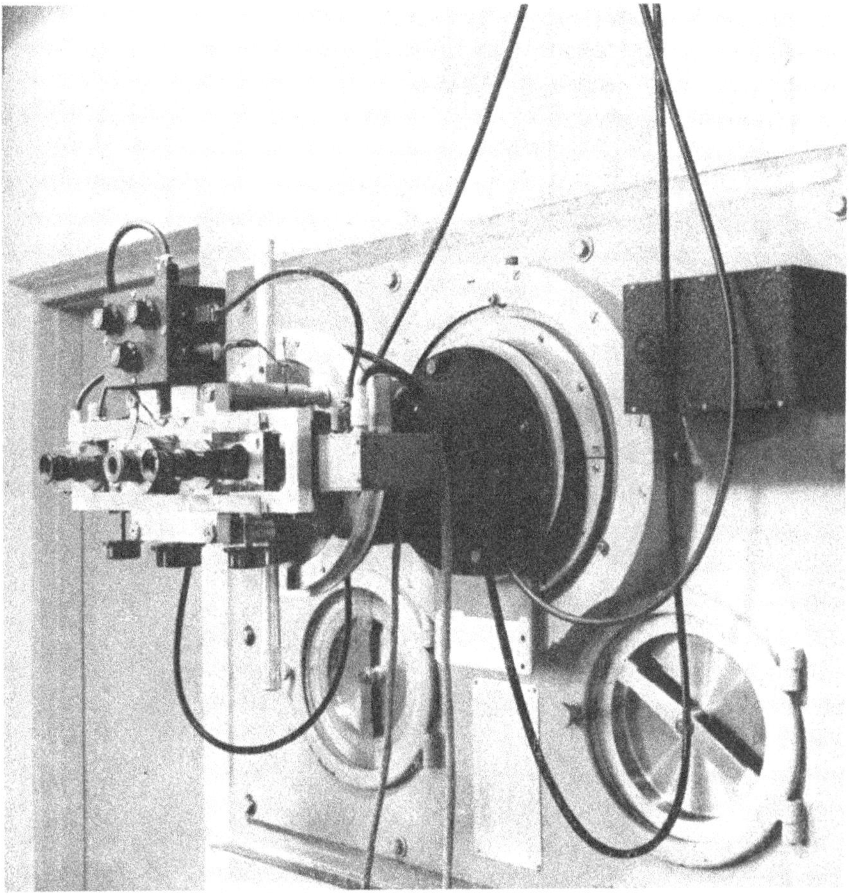

Plate VIII. The Pierce photo-electric photometer attached to the 15-inch horizontal telescope at Flower and Cook Observatory. The photometer allows simultaneous observation of the variable star and the comparison star. (Courtesy Dr. W. Blitzstein.)

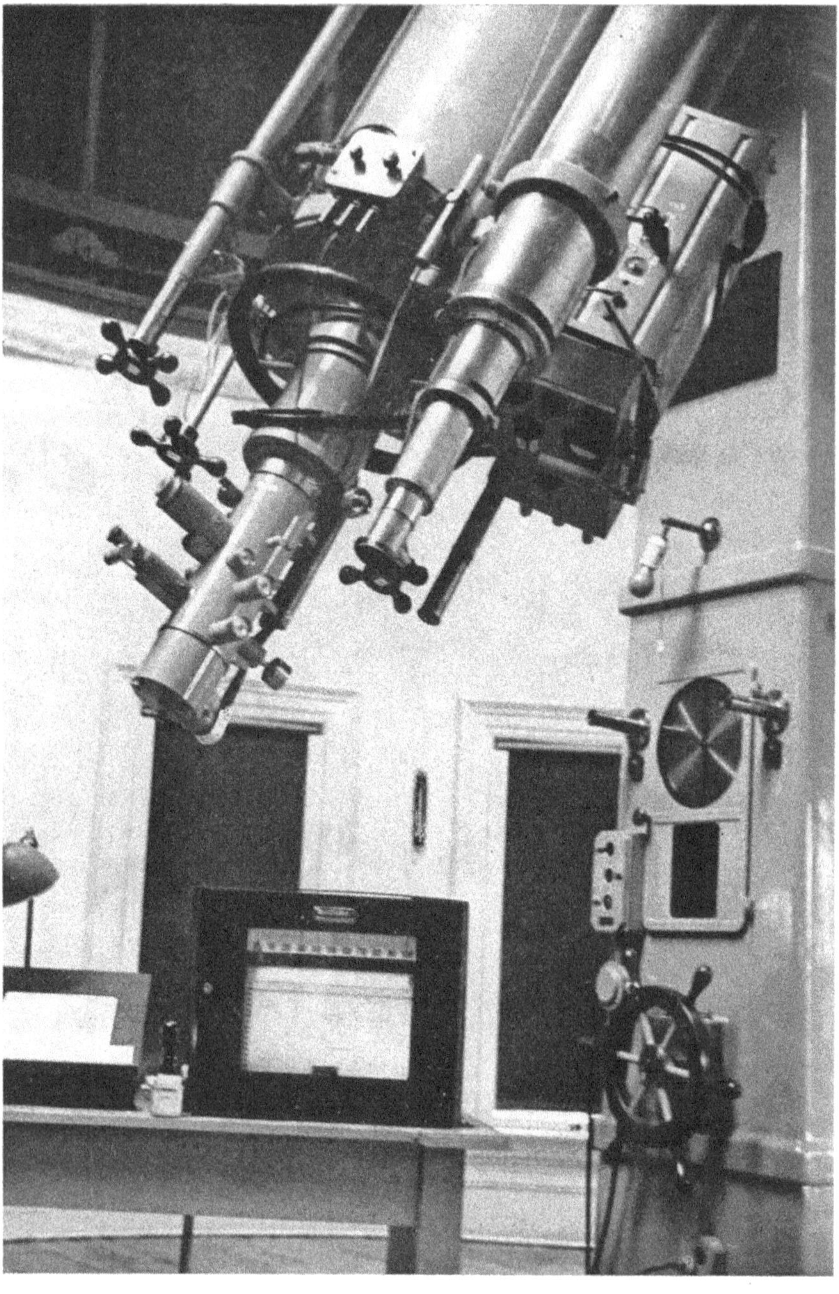

Plate IX. Photo-electric photometer attached to the 16-inch Goodsell refractor, now attached to the 28½-inch reflecting telescope at Flower and Cook Observatory. With this photometer one can observe the intensity, color, and polarization of star light. (Designed by the author.)

Most variables are discovered photographically. Several plates of the same region of the sky are taken over a period of time. A wide angle lens of short focal length is used, so that one plate covers a large part of the sky and reaches reasonably faint stars. In this way with the larger instruments, plates show about half a million stars in the rich fields of the Milky Way. It is clear that we have to inspect the plates taken at different times, but we cannot compare each star individually because it would take too much time.

The instrument that helps us is the blink comparator, or blink microscope, which unites two images of the same star in one microscope (Figure 114). If first the left image is illuminated

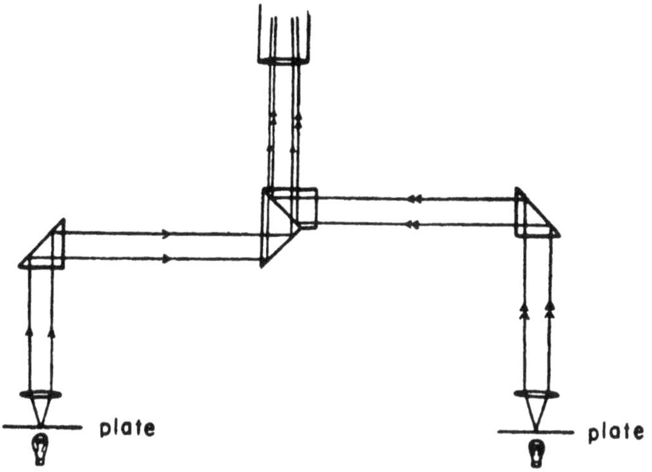

Figure 114. Principle of the blink microscope to detect variable stars and stars with large proper motion.

and then the right image, and so on in rapid succession, a star of constant diameter appearing in the microscope indicates that the light of the star remained constant on the two occasions. If, however, the star image in the microscope appears to pulsate, it indicates that the light changed. The same instrument can be used to detect a star of large proper motion, because then the star image appears to jump. These latter stars are usually nearby and can be placed on a program to determine their trigonometric parallaxes.

83. *Bright variable.* This is a relative method because we compare the variable's light with that of one comparison star of constant light. It is a good thing to compare this comparison star occasionally with another one, to see if the light really remains constant. Here we must know only the differential extinction, and if the two stars are but a small angular distance apart, this is negligible compared with the accuracy of the photographic magnitudes.

We take a plate with a sequence of grating exposures and thus simultaneously obtain three images of the comparison star and three of the variable. For each exposure we write down the previously defined table magnitude S for the comparison star and the variable, both for the central image and the mean of the first order images, a total of four magnitudes. These table magnitudes run practically linearly with the true magnitudes m, but because the relation is not perfect we adopt an additional quadratic term:

$$S = a + bm + cm^2 \tag{246}$$

We assign the subscript zero to the table magnitude of the central image, and one to the first order images. For the central image the true magnitude is m; for the first order image it is $(m + G_1)$ if G_1 is the grating constant for grating one. We wish to determine the true magnitude difference between the variable v and the comparison star c, that is $\Delta m = m_v - m_c$. In the computation the constants a, b, c, which are of no interest, will drop out.

$$
\begin{aligned}
S_{v0} &= a + b\,m_v \qquad\qquad + c\,m_v^2\\
S_{v1} &= a + b(m_v + G_1) + c(m_v + G_1)^2\\
S_{c0} &= a + b\,m_c \qquad\qquad + c\,m_c^2\\
S_{c1} &= a + b(m_c + G_1) + c\,(m_c + G_1)^2
\end{aligned}
$$

By subtraction we get:

$$
\begin{aligned}
S_{v0} - S_{c0} &= b\,\Delta m + c\,\Delta m(m_v + m_c)\\
S_{v1} - S_{c1} &= b\,\Delta m + c\,\Delta m(m_v + m_c + 2G_1)\\
S_{v0} - S_{v1} &= -\,bG_1 - cG_1(2m_v + G_1)\\
S_{c0} - S_{c1} &= -\,bG_1 - cG_1(2m_c + G_1)
\end{aligned}
$$

From this we find:

$$
\begin{aligned}
S_{v0} - S_{c0} + S_{v1} - S_{c1} &= +\,\Delta m\,\{2b + 2c(m_v + m_c + G_1)\}\\
S_{v0} - S_{v1} + S_{c0} - S_{c1} &= -\,G_1\,\{2b + 2c(m_v + m_c + G_1)\}
\end{aligned}
$$

By division we find Δm:

$$
\Delta m = G_1 \frac{(S_{v1} + S_{v0}) - (S_{c1} + S_{c0})}{(S_{v1} - S_{v0}) + (S_{c1} - S_{c0})} \tag{247}
$$

84. *Faint variable.* The scale of true magnitudes m is derived from a separate set of about ten grating plates according to the procedure for a rich star field. Because the variable is so faint, it takes too much time to pursue the regular observing sequence of using a grating in this way. In the actual observing program we take a sequence of exposures without a grating, from which we can get the table magnitudes S. Here we shall use about five comparison stars covering the entire magnitude range of the variable. A linear relation between the table magnitudes S and the true magnitude m for the comparison stars is postulated, thus $S = a + bm$. This can be checked with a graph if necessary. Entering with the known table magnitude for the variable, we arrive at its true magnitude. This procedure is done by least squares for each exposure. The probable error of one such relative magnitude determination is about $\pm$ 0.05 magnitude.

Another method is to make density marks on the plate directly after observation. The light ratios used are known, and the observed densities enable us to construct a reduction curve. This is often done for photometry on spectral plates.

85. *Light time*. The time is the most accurately observed quantity. At our place of observation we observe a certain signal of a star, for example, the minimum. We first convert the time to Greenwich Standard Time or Universal Time, which is the local time at which an observer at Greenwich would observe the signal. These hours, minutes and seconds are converted into decimals of a day. The day itself we denote by a number because this arrangement is so much simpler. This is the so-called Julian Day, counted from 1 January 4713 B.C., and given in the almanac for any date. By international agreement the Julian Day begins at Greenwich noon, so that for most active observers the night falls in the same Julian Day.

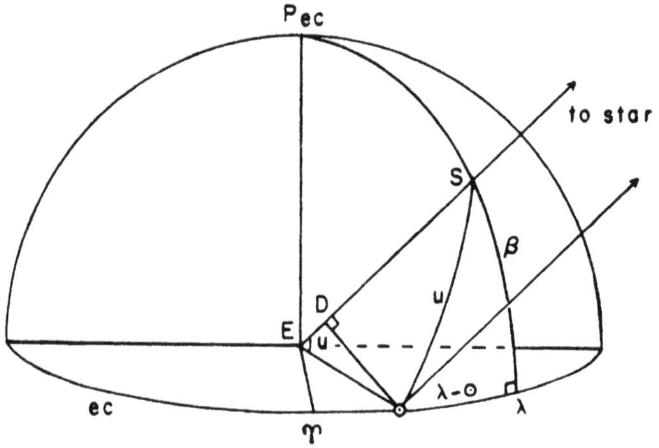

Figure 115. The light time is the time needed for the light to travel the distance DE.

This geocentric time still has to be converted to heliocentric. From a star in the ecliptic a signal may reach the earth between eight minutes earlier and eight minutes later than it reaches the sun, due to the finite velocity of light. In ecliptic coordinates we have in Figure 115:

$$\cos u = \cos \beta \cos (\lambda - \odot)$$

where λ and β are the celestial longitude and latitude of the star, $\odot$ is the longitude of the sun, and u is the angular distance between the star and the sun. The light time correction is the time needed for light to travel the distance DE in the figure. It travels an astronomical unit in $0\overset{d}{.}005770$, so that we have:

$$\Delta t = -0\overset{d}{.}005770 \cos \beta \cos (\lambda - \odot) \qquad (248)$$

This is correct for the circular orbit of the earth. Let us now find this correction in the equatorial system. In $\Delta\ SP_{eq}P_{ec}$ we have in Figure 116:

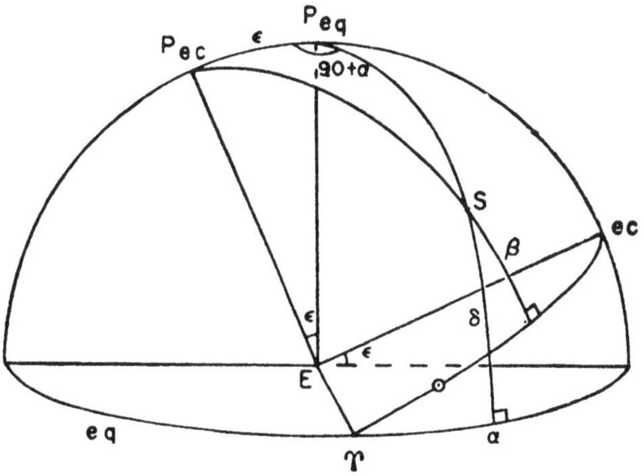

Figure 116. The parallactic triangle $SP_{eq}P_{ec}$.

$$\left.\begin{aligned} \sin\beta &= \cos\epsilon\sin\delta - \sin\epsilon\cos\delta\sin\alpha \\ a = \cos\beta\cos\lambda &= \cos\delta\cos\alpha \\ b = \cos\beta\sin\lambda &= \sin\epsilon\sin\delta + \cos\epsilon\cos\delta\sin\alpha \end{aligned}\right\} \qquad (249)$$

Thus we find:

$$\cos u = \cos\beta\cos\lambda\cos\odot + \cos\beta\sin\lambda\sin\odot$$
$$= (\cos\delta\cos\alpha)\cos\odot + (\sin\epsilon\sin\delta + \cos\epsilon\cos\delta\sin\alpha)\sin\odot$$
$$\cos u = a\cos\odot + b\sin\odot$$

Another way to find cos u is by using the cosine rule in the equatorial system (Figure 117).

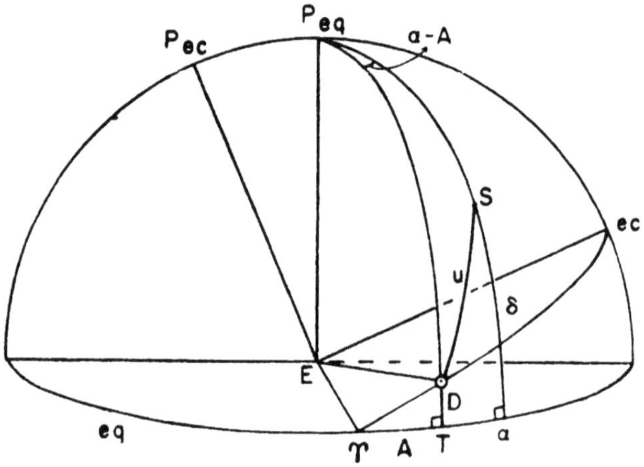

Figure 117. The spherical triangle $\odot SP_{eq}$.

$$\cos u = \sin\delta\sin D + \cos\delta\cos D\cos(\alpha - A)$$
$$= \sin\delta\sin D + \cos\delta\cos D\cos\alpha\cos A + \cos\delta\cos D\sin\alpha\sin A$$

In $\Delta \Upsilon \odot T$ we have:

$$\left.\begin{aligned} \cos\odot &= \cos D\cos A \\ \sin\odot\sin\epsilon &= \sin D \\ \sin\odot\cos\epsilon &= \cos D\sin A \end{aligned}\right\} \qquad (250)$$

By substitution we find:

$$\cos u = \sin \delta \sin \odot \sin \epsilon + \cos \delta \cos \alpha \cos \odot + \cos \delta \sin \alpha \sin \odot \cos \epsilon$$

$$= (\cos \delta \cos \alpha) \cos \odot + (\sin \epsilon \sin \delta + \cos \epsilon \cos \delta \sin \alpha) \sin \odot$$

$$\cos u = a \cos \odot + b \sin \odot$$

The light time in a circular orbit is thus:

$$\Delta t = 0^{d}.005770 \, (a \cos \odot + b \sin \odot) \qquad (251)$$

The almanac gives the projection of the sun on the equatorial plane, where R is the radius vector of the earth for a given date (Figure 118).

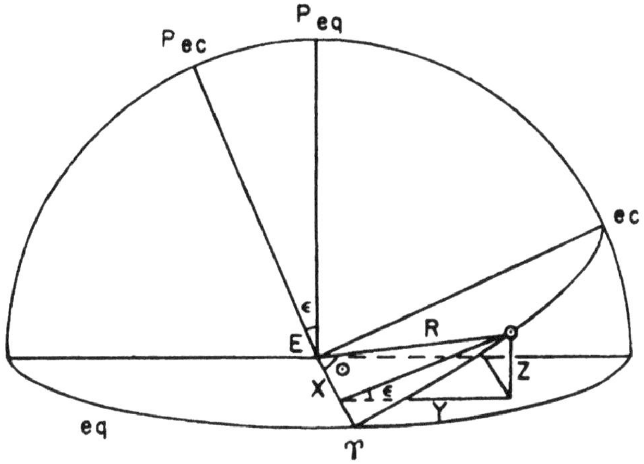

Figure 118. Projection of the sun on the equatorial plane.

$$\left. \begin{aligned} X &= R \cos \odot \\ Y &= R \sin \odot \cos \epsilon \end{aligned} \right\} \qquad (252)$$

The expression for $\cos u$ becomes:

$$\cos u = (\cos \delta \cos \alpha) \cos \odot + (\tan \epsilon \sin \delta + \cos \delta \sin \alpha) \cos \epsilon \sin \odot$$

The light time in an elliptical orbit is now:

$$\Delta t = -0^{\text{d}}.005770\, R \cos u$$
$$= -0^{\text{d}}.005770 \{(\cos\delta\cos\alpha)X + (\tan\epsilon\sin\delta + \cos\delta\sin\alpha)Y\}$$
$$\Delta t = -0^{\text{d}}.005770\,(A'X + B'Y)$$

Usually this is written in the form:

$$\Delta t = AX + BY \tag{253}$$

This is a distorted cosine curve. The correction has to be applied to make the time heliocentric, as seen for an observer on the sun. Here we can make a check using a table by R. Prager. No correction is made for the linear solar motion.

It is possible to construct a clock which gives four decimals of a day. The light time for the star under observation is a known zero point correction for the date of observation and can be taken into account beforehand, and the heliocentric time can thus be recorded directly in decimals of a day.

Since the stars are so far away, the magnitudes are the same whether observed from the sun or from the earth. This is not so if the light curves of the minor planets or asteroids are observed.

86. *Period.* If the light variation is periodic, we should like to find the period. We usually do not know the type of variation beforehand. The best way to begin is to collect the fainter magnitudes. If we find several provisional minima in this way, we know that their separations in time must be a whole number of periods. Let a be the epoch or the Julian Day of the first minimum, the smallest time interval as the period P in days, and n the number of periods elapsed since our initial one. We can write then, if t_{m} is in Julian days:

$$t_{\text{m}} = a + Pn \tag{254}$$

For a given observation at time t we compute now $(t - t_{\text{m}})$ or the time interval elapsed since the last minimum occurred. We define now the phase as the fraction of the period counted between 0 and 1, and find:

$$\text{Phase} = \frac{t - t_{\mathrm{m}}}{P} = (t - t_{\mathrm{m}})P^{-1} \qquad (255)$$

We compute phases for all known times of observation. On the computing machine this is done by multiplying by the reciprocal of the period. In other words we shift all observations a certain number of periods and bring them together to form one light curve. Now we can make a sharper determination of the period using the middle of ascending or descending branches. In the case of an eclipsing variable the minimum is usually symmetrical and the temporal mean of both branches determines the minimum time much more precisely than does the actual observed minimum time. An accurate and impersonal method of computing this time is given by E. Hertzsprung. In practice the period is then determined by least squares from the above mentioned formula (254) where a and P must be determined. To get the probable errors in the correct way, the epoch has to be taken approximately in the middle of the time interval of observation. From the definite period we compute definite phases and determine the normal points of the light curve. In the case of an eclipsing system the phase is thus made zero at primary minimum. A well determined period is a necessity before one can start with orbital solution.

87. *Photo-electric method.* When light energy strikes certain metals, electrons can be emitted from the metal under proper circumstances. The number of electrons emitted is proportional to the light intensity, but their velocity is independent of this intensity and depends on the frequency or on the wavelength of the light. In the old-fashioned photo-electric cell the star light falls on a thin layer of an alkali metal (K, Li, Na, Rb,

Cs), which has been sealed inside a glass bulb filled with an inert gas. This plate is called the cathode. The electrons move to the positively charged anode, producing a measurable flow of electrons from the plate to the anode.

A typical multiplier photocell (1P21) is about the size of a small radio tube. Light through the window falls on the cathode 0 from which electrons move in a curved orbit to plate 1, from which 3 to 6 electrons are set free. The curved orbits are caused by the 90 volt difference between successive dynodes. The electrons now strike plate 2, and so on, until a current amplification of about 1.5 million has been achieved (see Figure 119).

Under the most common observing procedure a linear amplifier is used to increase the current once more. With the

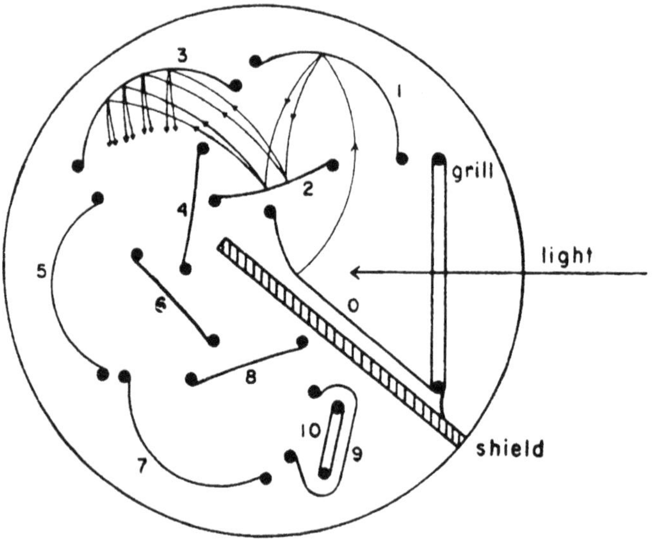

Figure 119. Schematic arrangement of type 1P21 multiplier photocell. 0 = cathode, 1 – 9 = dynodes, 10 = anode.

most sensitive step, we get again an amplification of a factor of one million. The total result is that we can measure the intensity of any star we can see through the telescope. The sensitivity of the instrument is thus the same as that of the eye, but its accuracy is much better.

The current must be recorded. Good instruments for this purpose are the Brown, Speedomax, or similar recorders. These are really potentiometers which record the voltage drop across a standard resistance through which the amplifier output is sent. A paper rolls on a cylinder at a speed of one inch per two minutes, for example, and at the same time a pen records the deflection. In this way we keep a record of the observation. Moreover, if something goes wrong during the observation, it shows up immediately, and can be corrected at once.

Another technique is the pulse counting method, in which we measure the number of anode pulses above a certain level. This number, which can be printed on a roll of paper, is linear with the light intensity.

In the optical path a diaphragm must be placed in the focal plane of the objective or mirror. The star is fixed in this small diaphragm and the largest part of the sky is thus cut off. In the case of a mirror the rays of all wavelengths converge to one point, so that the diaphragm can be made small. In the case of a refractor the visual light may be in focus in the diaphragm, but the ultraviolet light will be out of focus and thus give a large and diffuse image in the plane of the diaphragm. It is best to focus in the blue region and make the diaphragm larger, so that the light of all wavelengths is allowed to enter through the diaphragm. Then different color filters are customarily used in the light path so that light of a known color is transmitted to the photocell. This light is outside the focus when it falls upon the cathode (Figure 120).

The clock drive may not be excellent and the star image may

shift somewhat in the focal plane; bad seeing may also produce the same effect. The light would not hit the same spot on the cathode of the cell, and because the sensitivity of the cathode may change from point to point, this could be a serious objection. An additional difficulty in a 1P21 multiplier photocell is

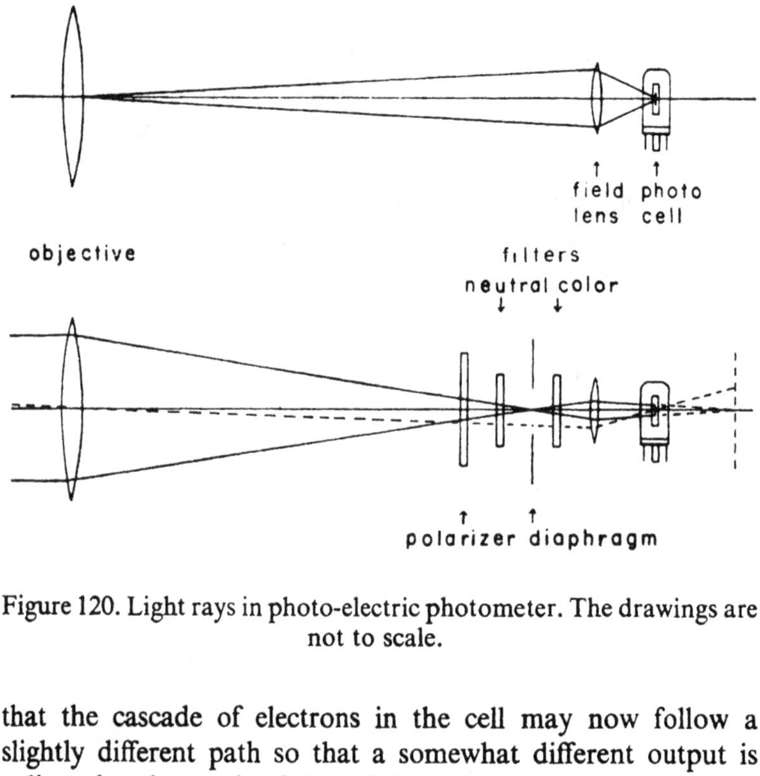

Figure 120. Light rays in photo-electric photometer. The drawings are not to scale.

that the cascade of electrons in the cell may now follow a slightly different path so that a somewhat different output is collected at the anode of the cell. Therefore, a positive field lens is placed in the light path to give a sharp image of the objective or mirror on the cathode, and thus remove the trouble. The light spot on the cathode remains steady even if the star image moves in the focal plane. We can say that the photocell measures the brightness of the objective or mirror as illuminated by the star.

A polarizer may be placed in the light path; by rotation of the polarizer the amount and plane of polarization of the star light can be measured.

We have already described how the transparency curve of a filter can be found. The sensitivity of the cell can be found by comparing the light intensity on the cell with the same light intensity on a thermocouple or bolometer.

A linearity test of the recorder can be made by introducing a very small electrical current into the instrument. If a second current is made a multiple of the first one, the ratio of deflections will show any deviation from linearity. The same procedure can be done with the combination amplifier-recorder. Such linearity is necessary, and if there are deviations from it we must determine these discrepancies. We usually have $I = cd$, where for deflection d, the constant c is determined as soon as the unit of intensity I has been fixed. We can then fix the current and thus the deflection 2.512 times the previous value, corresponding to an increase of one magnitude. This means only that a factor of one hundred in intensity must correspond to five magnitudes. In other words, we can calibrate the instrument according to the Pogson scale, and need the North Polar Sequence only for the determination of the zero point.

88. *Open clusters.* For photometry of open clusters or other rich star fields we make absolute observations. Excellent transparency is necessary, and it must remain constant all night. During a particular night the extinction must be determined and the extinction corrections applied. The magnitudes are then reduced to their values outside atmosphere; independent of the atmospheric conditions of the various places of observation, all observers should get the same results within the probable errors. This is what is meant by an absolute result; consequently not all

stars are measured absolutely and independently of the others. We will measure the stars in an open cluster in a relative way, namely with respect to a small number of standard stars, so that the group as a whole is homogeneous.

The observations of one star may proceed as follows: dark current, star plus sky in yellow, star plus sky in blue, sky yellow, sky blue; the sequence is then repeated in reverse. The dark current is measured to see whether it has remained constant during the observation. The sky light may change rather quickly when the moon is rising or setting. By subtraction we easily find star yellow, star blue, and thus the relative deflections. For extinction stars we choose both a red and a blue one which are low in the east or northeast at the beginning of the night's work. If the open cluster is itself situated there, we can often take these extinction stars as standard stars and save some observational work. The sequence for stars of the open cluster may be arranged as follows: radium light, red standard star, blue standard, star 1, star 2, red standard, blue standard, star 3, star 4, radium light. A certain star in the cluster is consequently always closely timed between two standard stars, and the extinction is frequently determined.

The relation between magnitude and deflection is:

$$m = -2.5 \log l = -2.5 \log cd = -2.5 \log c - 2.5 \log d$$

The proportionality factor c thus enters as a term in the magnitude. Provisionally we can take $c = 0.01$ when the scale is given between 0 and 100, with the advantage that the provisional magnitudes are small positive numbers. The final zero point determination of the magnitudes fixes the final value of the intensity unit, and thus also the final value of the constant c.

For relative comparison between two stars the constant c drops out as long as the relation is linear. For the cluster star and the standard star we have for example:

$$\Delta m = -2.5 \log \frac{l}{l_s} = -2.5 \log \frac{d}{d_s} \qquad (256)$$

Only the ratio of the deflections is used. In this way we are sure that the scale is correct as soon as the scale of the standard stars is correct, and these standard stars are measured on many different nights. The magnitudes found in this way are those for the effective wavelength of the filter-cell combination and contain a slight dependence on spectral type.

By measuring blue and yellow magnitudes, one immediately after the other, we find the color according to:

$$C = m_b - m_y = -2.5 \log \frac{l_b}{l_y} = -2.5 \log \frac{d_b}{d_y} \qquad (257)$$

Here we measure only two deflections and compute the ratio. The color thus found is determined by the two effective wavelengths or the color base line. This color determination is highly superior to the photographic one, because the zero points and scales of both magnitude systems do not enter here, but only the linear relation. Both magnitude and color are found in the natural system corresponding to the effective wavelengths of the combination.

89. *Extinction*. In the atmosphere we can imagine parallel layers of air; we have a similar relation as in paragraph 75.

$$\frac{l_y}{l_0} = e^{-\text{air mass}}, \qquad \Delta m = K \times \text{air mass} \qquad (258)$$

For all layers together this still holds. If we take as unit the air mass which star light has to traverse in the zenith, the air mass in the direction of zenith distance z equals sec z. We shall first compute sec z as a function of the hour angle τ, which we can easily observe on the telescope, or compute from the sidereal time and the right ascension of the star. With the cosine rule in $\Delta P_{eq}ZS$ we have for latitude φ of the observer (Figure 121):

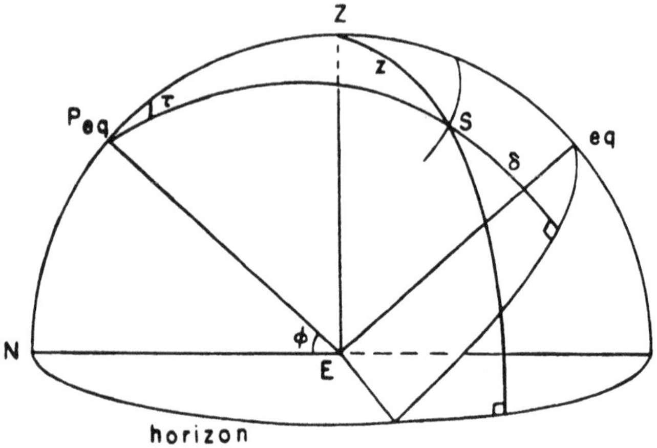

Figure 121. The spherical triangle $P_{eq}ZS$ to determine the zenith distance and the air mass.

$$\cos z = \sin \varphi \sin \delta + \cos \varphi \cos \delta \cos \tau$$

$$\sec z = \frac{1}{\cos z} = \frac{1}{\sin \varphi \sin \delta + \cos \varphi \cos \delta \cos \tau} \tag{259}$$

For a certain place of observation φ is constant. For all stars of the same δ the air mass can be computed as a function of τ. An extinction curve for every 5° declination simplifies the computation. We must observe in an orderly fashion, and need only record the time and hour angle at the beginning. Later on, during the reduction, the other hour angles can easily be found.

In the following we will take all extinction coefficients (K) positive. Let m' and sec z be the observed quantities and m the real magnitude outside the atmosphere, then we have thus in Figure 122:

$$\left.\begin{aligned} m'_b &= m_b + K_b \sec z \\ m'_y &= m_y + K_y \sec z \end{aligned}\right\} \tag{260}$$

The respective extinctions in the zenith, for $z = 0$ or $\sec z = 1$, are therefore K_b and K_y in the blue and the yellow. For the

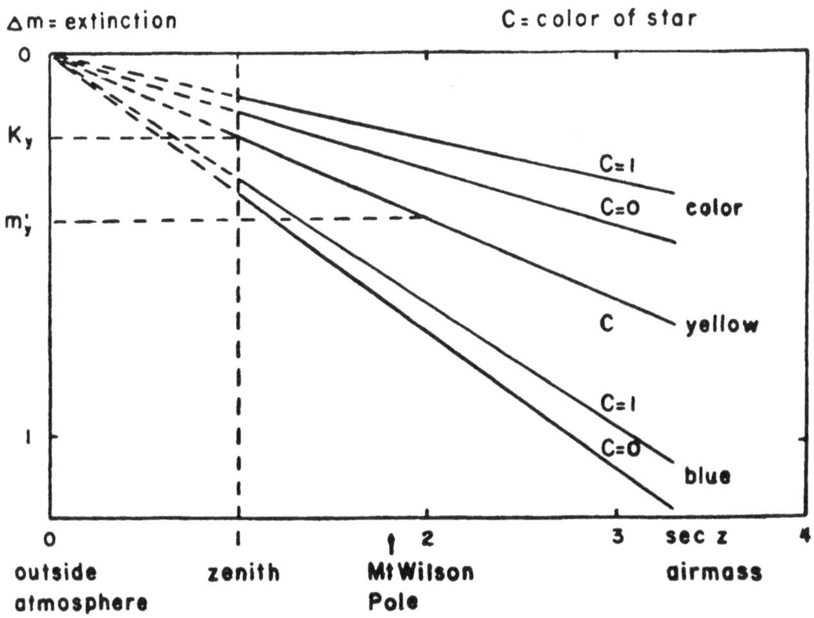

Figure 122. The extinction plotted against air mass. The extinction coefficient for yellow wavelength is practically independent of the color of the star.

colors we observe $C' = m'_b - m'_y$ and define $C = m_b - m_y$ as the color outside the atmosphere. Let $K_c = K_b - K_y$ be the color extinction in the zenith.

$$C' = C + (K_b - K_y) \sec z = C + K_c \sec z \qquad (261)$$

For one night's observation we can find m_b and K_b, m_y and K_y, C and K_c for the individual extinction stars. It is found now that both the color and the magnitude extinction are linear functions of the outside color of the star in the following way:

$$K_c = K_1 - K_2 C \qquad (262)$$

The color outside the atmosphere for any star observed during the night can now be found:

$$C' = C + (K_1 - K_2 C)\sec z$$
$$C = \frac{C' - K_1 \sec z}{1 - K_2 \sec z} \qquad (263)$$

The extinction must be observed on both sides of the meridian and should stay constant during the night. The extinction stars have to be observed at suitable time intervals so that the points on the extinction line are about equally spaced. This demands many observations when the extinction stars are close to the horizon, and only a few when they are close to the zenith.

In the same way one finds the relations for the magnitudes:

$$\left.\begin{aligned} K_b &= K_{b1} - K_{b2}C \\ K_y &= K_{y1} - K_{y2}C \end{aligned}\right\} \qquad (264)$$

K_{y2} is very small so that the yellow extinction coefficient is practically independent of the color of the star. It is of advantage, therefore, to determine m_y and the color. After determining these constants we find for any star:

$$\left.\begin{aligned} m'_b &= m_b + (K_{b1} - K_{b2}C)\sec z \\ m'_y &= m_y + (K_{y1} - K_{y2}C)\sec z \end{aligned}\right\} \qquad (265)$$

The same observations determine the result for both C and m. However, C has been determined relatively and m absolutely. This last determination can easily introduce systematic errors.

If more than one night's observations are available, we can improve our results considerably by correcting for these systematic errors. For every night the colors and magnitudes outside the atmosphere must be the same for the two or more extinction stars if these stars are not variable. The individual colors outside the atmosphere for a given extinction star are averaged and the same is done for the magnitudes. Thus $\overline{C}$ and $\overline{m}_y$ are computed for each extinction star from all observations. For the determination of $\overline{m}_y$ the observations of the standard radium source must be used to adjust the overall sensitivity of the equipment for each night to some mean value. The computation of $\overline{m}_y$ has physical significance only if this sensitivity adjustment is made. The colors and magnitudes outside the atmosphere are thus fixed, and the linear relation is forced through them:

$$\left.\begin{aligned} C' &= \overline{C} + K_c \sec z \\ m'_y &= \overline{m}_y + K_y \sec z \end{aligned}\right\}$$

K_c and K_y are next determined for each night by the method of least squares. With these newly computed K_c and K_y the K_1, K_2 and K_{y1}, K_{y2} are computed for each night from:

$$\left.\begin{aligned} K_c &= K_1 - K_2\overline{C} \\ K_y &= K_{y1} - K_{y2}\overline{C} \end{aligned}\right\}$$

This gives the best possible extinction coefficients. For any star we compute the best natural color and magnitude according to formulae (263) and (265). The probable error of one color determination is of the order of $\pm$ 0.005 magnitude, and for a magnitude determination of the order of $\pm$ 0.01 magnitude. In the

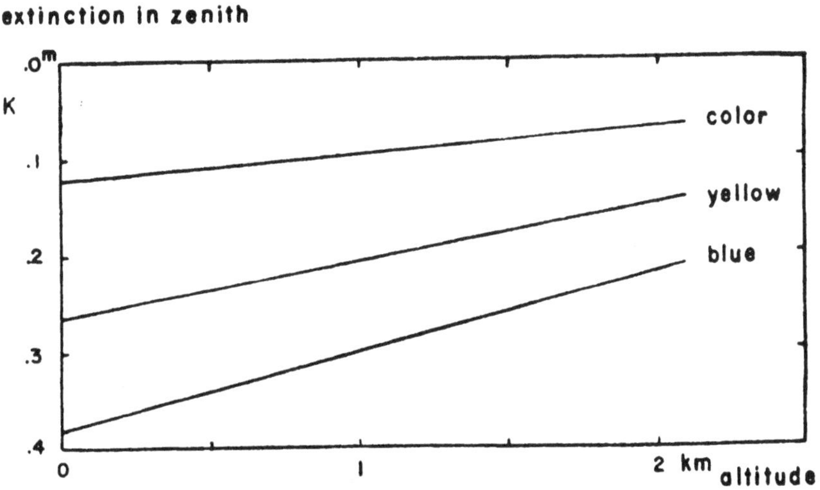

Figure 123. On good clear nights the extinction in the zenith for stars of averaged color depends on the altitude of the observatory.

latter probable error, the extrapolation to outside atmosphere is included.

A linear formula will convert these natural colors and magnitudes to the international ones. For the magnitudes we have, in addition, the color coefficient, due to the comparison of the magnitude sequences in two effective wavelengths.

On good clear nights $\overline{K}_y$ depends on the altitude of the place of observation, since K_y is the extinction in the zenith. For an observatory on a mountain there is less atmosphere above the observer, and therefore K_y will be small (Figure 123). There is indeed a linear relation between K_y and the altitude. For K_b and K_c stars of the same color must be used to find similar relations.

90. *Light curve*. Observations of a variable are made relative to a comparison star; therefore it is easier to make them with precision. Some photometers are especially constructed for this field of photometry and are made to observe the variable and the comparison star simultaneously. The influence of the change in transparency becomes then very much smaller in the relative measurements. In one such photometer the light of the variable and the comparison star falls on separate cells and the amplifiers are in duplicate. There is an arrangement for interchanging cells and amplifiers for checking purposes (Plate VIII).

An ingenious technique devised by Th. Walraven makes use of a rapidly oscillating diaphragm to throw the light first of the variable and then of the comparison star on the same cell. The method makes use of a wedge and records directly the difference in magnitude between the two stars.

With an ordinary photometer not especially constructed for variables we must observe the variable and the comparison star alternately. The variable has to be timed between the two observations of the comparison star. The average value of the deflections of the comparison star is then computed at the same time as the observations of the variable, and the two can be compared. A regular observing schedule is necessary.

One can observe for example in this way: dark current, comparison star plus sky in yellow, comparison star plus sky in blue, sky blue, sky yellow, variable plus sky in yellow, variable plus sky in blue, sky blue, sky yellow, dark current. Occasionally we also observe the standard source and a second comparison star to see whether the light of the first comparison star really remains constant.

We can find thus the deflection d caused by the star light. The color of the star and the magnitude difference between the variable and comparison star are found as before.

$$C = m_b - m_y = -2.5 \log \frac{l_b}{l_y} = -2.5 \log \frac{d_b}{d_y}$$

$$\Delta m = -2.5 \log \frac{l_v}{l_c} = -2.5 \log \frac{d_v}{d_c} \qquad (266)$$

Here a graph or table giving the relation between m and l or d is very handy. It should be noted that both cases are relative measurements. No knowledge of the magnitude scale is necessary, since only the linearity of the instrument is important. Again C and Δm are the natural color and magnitude difference for the effective wavelengths. If we desire, we can make the conversion into the international color and magnitude.

The results must be corrected for relative extinction when the variable and the comparison star do not have a small angular distance in the sky or do have very different spectral types. In general we have $(K \sec z)_c - (K \sec z)_v$, which becomes for yellow magnitudes $K_y(\sec z_c - \sec z_v)$. A table or graph between the relative air masses and the hour angle of the variable can be profitably used. The observations of the comparison star give information about the extinction coefficient for the particular night. With some practice the yellow coefficient can be estimated by the naked eye just by looking at the number of stars visible.

91. *Polarization.* This polarization of star light is caused by interstellar grains. We first observe a nearby star where we expect no polarization. Any amount of polarization found must then be caused by the lens or mirror and can be corrected in the later reduction. The provisional measurement for a distant star can proceed as follows. Take observations at position angles 0°, 10°, 20°, ... 140° in rapid succession. Light from a star is only partially polarized, usually not more than about 10%. The variable part of the intensity will change according to $\cos^2(\theta - \theta_0)$ or $\cos 2(\theta - \theta_0)$ if we change the position angle θ of the polarizer. The maximum and minimum deflections are at 90° to each other and determine the position angle θ_0 of the plane of polarization. The amount of polarization can also be given as a difference in magnitude from the maximum and minimum deflections (Plate IX).

In practice the following procedure gives more accurate results. Three deflections are made, the first and third at the same position of the polarizer and the second one with the polarizer rotated by 90° from the other positions. The magnitude difference can be determined. This process is repeated for other position angles. If polarization is present then a $\cos 2(\theta - \theta_0)$ curve should result. The amplitude gives the amount of polarization and the position angles of the extreme values give the plane of polarization. This plane defined by the electrical vector and the line of sight, is in general parallel with the Milky Way plane, if this line of sight is not in the direction of a spiral arm.

Sunlight and moonlight are polarized; therefore we cannot observe with the moon in the sky. The daytime sky shows the polarization to depend both on the position of the sun, the maximum being in a large circle 90° from the sun, and on the horizon of the observer.

Planets show polarization depending on their longitude difference from the sun, proving that they reflect sunlight. Polarization of planetary light tells something about the conditions on the surface. The moon and Mercury show similar polarization curves.

W. A. Hiltner gained another factor of 10 in accuracy by observing both the ordinary and extraordinary beams of star light passing through an Iceland crystal. He works with separate photocells and amplifiers.

92. *Moving cluster parallax.* This parallax can be determined for nearby clusters like the Hyades, the Ursa Major group, and the Scorpius-Centaurus group. The members of these groups have parallel motions in space, which by perspective we observe to converge or to diverge from a certain point on the sphere, the so-called convergent point. This corresponds to the radiant in the case of a meteor shower. The convergent point is thus observed from the proper motions of all the cluster members. For

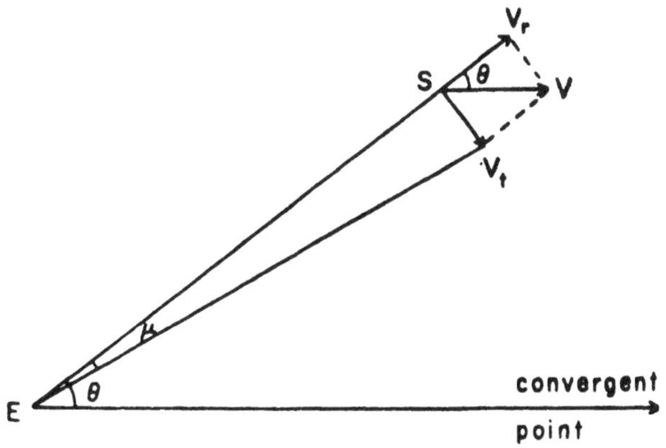

Figure 124. Convergent point for stars in the Hyades.

a certain member we can observe the angle θ, which is the angle between the space velocity and the line of sight. It is measured as the angular distance between the star and convergent point. Further, we measure the proper motion μ and the radial velocity V_r. For an individual member we then have in Figure 124:

$$V_r = V\cos\theta, \qquad V_t = V\sin\theta$$

$$V = \frac{V_r}{\cos\theta}$$

We can thus find the space velocity V and then V_t in kilometers per second. One astronomical unit per year corresponds to 4.74 km/sec. If we express the proper motion μ in seconds of arc per year we have $4.74\,\mu = V_t/r$. For the parallax of a member star it follows:

$$\pi_y = \frac{1}{r} = \frac{4.74\,\mu}{V_t} = \frac{4.74\ \mu}{V\sin\theta} \tag{267}$$

It is best to compute V separately for each member and then take the average value because the space velocity is very nearly the same for all members. Then compute π_y for each member, and take the average value as the mean parallax of the whole cluster. However, we shall certainly find that some stars are closer than the gravity center and others more distant. This affects the magnitudes, and we apply a correction to bring them to the average distance. This parallax method is a geometrical one just as is the trigonometric parallax, and thus it is not influenced by interstellar absorption. Other methods in which absolute magnitudes enter are so affected, and have to be corrected for such absorption.

93. *Spectrum-magnitude diagram.* We can construct a spectrum-magnitude diagram, also called the Hertzsprung-Russell diagram after the authors who first published such a plot. For the Hyades, for example, we can easily make a $Sp - m$ diagram,

and because we know the distance of the group we can find the distance modulus $m - M$ and thus also plot the $Sp - M$ diagram. This is actually the same picture, but there is a zero point difference between m and M. As already stated, we use the parallaxes of the individual members. In this case only the cosmic scatter and observing errors enter into the diagram. The cosmic scatter is found to be small (Figure 125).

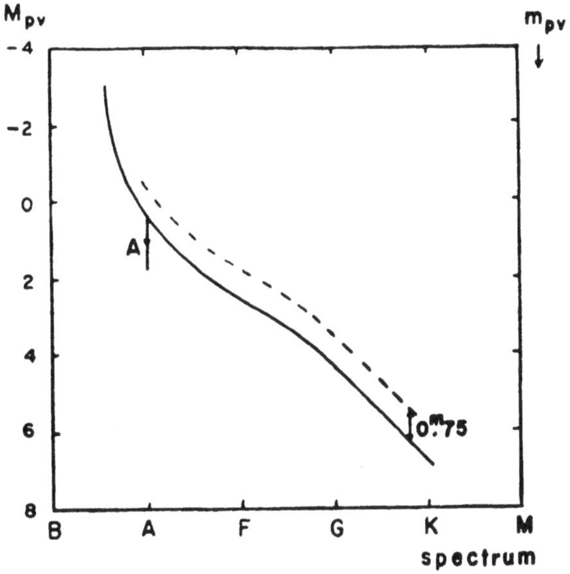

Figure 125. The main sequence of an open cluster in the spectrum-magnitude diagram. The double star branch is $0^m\!.75$ higher. A member hidden behind a dust cloud falls below the main sequence.

The same $Sp - M$ diagram can be plotted, using the nearby stars with well determined trigonometric parallaxes. This graph is a good example of the stellar distribution in the neighborhood of the sun, and it shows that there are many dwarf stars and only a few giants.

If one of the stars is hidden behind a dust cloud the spectral lines and therefore the spectral type remain the same. Interstellar absorption affects the ordinate, and so the star will fall below the main sequence by an amount equal to the absorption.

If there is a star in the main sequence that can be seen only singly, but in reality is a double, the observed point will be situated higher than the main sequence. Let the components be identical normal main sequence stars. Both are thus assumed to have the same intensity. The total effect is an intensity of 2, which corresponds to a magnitude difference $\Delta m = -2.5 \log 2 = -0^{m}.75$ brighter than the main sequence. We can call this the double star branch. This is a photometric way to discover unresolved doubles in open clusters.

94. *Color-magnitude diagram.* For the Hyades and the nearby stars with good trigonometric parallaxes we can construct the $C - m$ diagram and also the $C - M$ diagram. Again we use the individual parallaxes. The colors are measured photo-electrically and are quite accurate, so that this diagram gives a good idea of the small cosmic scatter involved. We shall again have a double star sequence (Figure 126).

A star hidden behind a dust cloud will have an interstellar absorption which brings the star point down in the vertical direction, and also a reddening which shifts the point to the right in the horizontal direction. The total effect is that the star will be shifted to another position on the main sequence. The total appearance of the main sequence therefore remains the same, except for the very early type stars. This diagram is therefore of no use in the discovery of reddening.

Let us now consider the Pleiades cluster, where half of the field is covered by interstellar material. This is partly visible on photographs; star counts show the absorption regions where a lack of field stars is apparent. Long exposures with the Mount

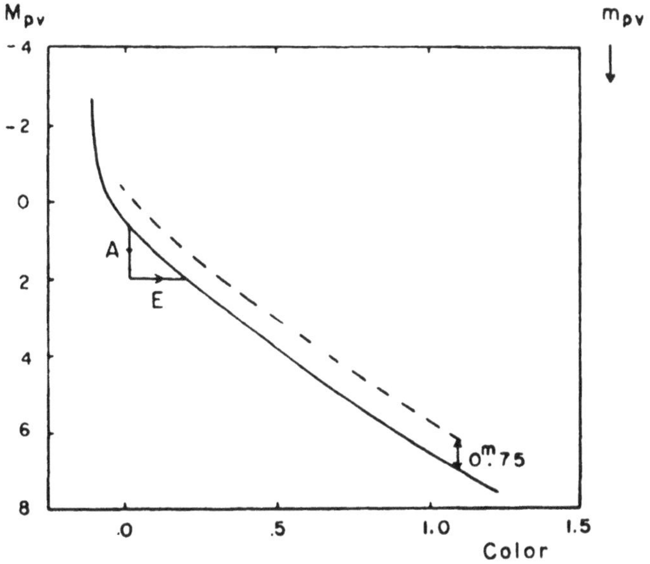

Figure 126. The main sequence of an open cluster in the color-magnitude diagram. The double star branch is $0^m.75$ higher. A member behind a dust cloud falls again on the main sequence.

Wilson telescopes show the extra-galactic nebulae shining through in some parts of the field, which we can call windows. The *Sp* − *m* diagram shows the absorption of the individual members. In the absorption-free parts of the field we can take the cluster members and plot their spectral types against the normal colors. This gives us a curved relation. If we now take a cluster member in the absorbed part of the field, we can plot its spectral type against the observed color in the same diagram. The points for these last cluster members are shifted so that the reddening can be read off. The ratio of absorption and reddening follows.

95. *Open cluster parallax.* The other open clusters are so

small in angular diameter that the convergent point cannot be determined. We can still make a $Sp - m$ diagram or a $C - m$ diagram and compare these diagrams with the above mentioned $Sp - M$ diagram or $C - M$ diagram. Thus for each cluster we can determine the distance modulus or the shift in magnitudes $m_0 - M = 5 \log r - 5$, where m_0 is the apparent magnitude corrected for absorption. We assume that all clusters have the same kind of main sequence, thus for example, that an *A*0V type star always has the same normal color and the same absolute

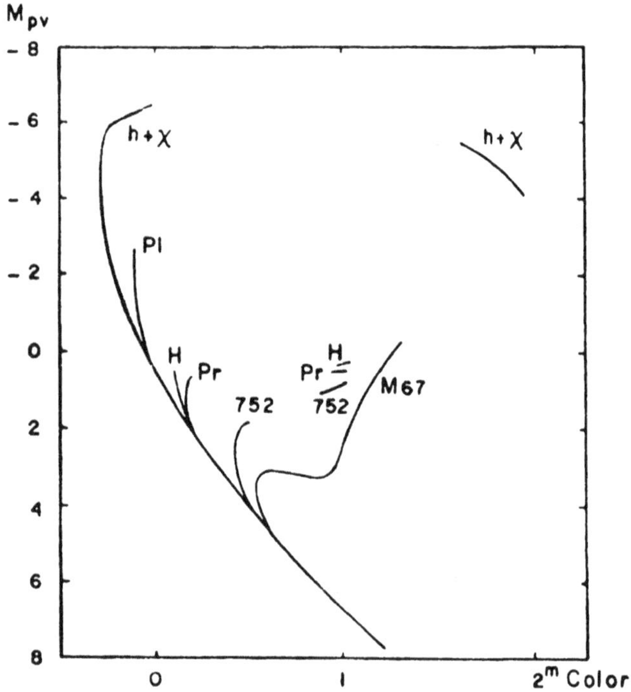

Figure 127. Comparison of the color-magnitude diagrams of the following open clusters: $h + \chi$ Persei, Pleiades, Hyades, Preasepe, NGC 752 and M67.

magnitude. This method is really a special case of the spectroscopic parallax in that we make the determination for all main sequence stars at the same time. We can say that we take a star of spectral type *A*0V of which we observe m and know M from geometrical methods; thus $m_o - M$ follows.

But our assumption is not necessarily true. Indeed, the cluster luminosity depends on the hydrogen content of the members of the cluster, and this may differ somewhat from one cluster to another, depending on the cluster age. In the youngest clusters the lower part of the main sequence is displaced upward or to the right. In the oldest clusters the upper portion of the main sequence is distorted and is quite different for different open clusters (Figure 127).

The open clusters can be divided into subgroups, which have about the same size in space, and thus their apparent diameters are a measure of their distance. On the other hand $m - M$ derived above must be corrected for absorption. If we fail to apply this correction, there is a disagreement for the distant clusters. This was the way in which R. J. Trumpler first found the general absorption and reddening in the Milky Way plane.

96. *Globular clusters.* There are about 100 known globular clusters centered around the nucleus of the Milky Way. The color-magnitude diagram of such a cluster is radically different from that of an open cluster (Figure 128). The distance can be determined with help of the cluster variables for which $M = 0$; thus $m - M$ and the distance follow. The cluster type variables are similar to the RR Lyrae type variables in our neighborhood for which we can determine this absolute parallax and absolute magnitude. We then find that the observed members of the globular cluster are all brighter than the sun and are mainly giants. Only the beginning of the main sequence is observed, and we would need still larger telescopes to observe the fainter stars.

The cluster variables are in a very special part of the horizontal branch, and obviously the star balance is upset there.

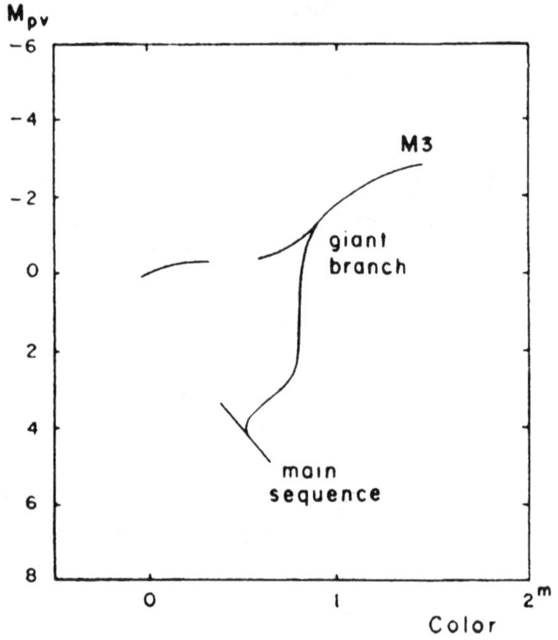

Figure 128. The color-magnitude diagram for a globular cluster, showing the giant branch and part of the main sequence.

This difference was one of the reasons why W. Baade introduced the concept of two populations. Population I is formed of stars in our solar neighborhood and thus comprises typical spiral arm stars. The stars belonging to the nucleus of the Milky Way form the Population II. The cluster stars and globular clusters are included in this population II. There are many differences between the two populations, of which the most important are the color-magnitude diagram, the space velocity, and the spectral behavior.

REFERENCES

BLACK BODY RADIATION

E. A. Milne: *Thermodynamics of the stars.* Handbuch der Astrophysik, Springer Verlag, Berlin, **3**, part 1, p. 65, 1930.

S. Chandrasekhar: *Stellar structure.* Un. of Chicago Press, Chicago, Chapter V, 1939

M. Waldmeier: *Einführung in die Astrophysik,* Birkhäuser Verlag, Basel, p. 17, 1948.

L. H. Aller: *Astrophysics, The atmospheres of the sun and stars.* Ronald Press Co., New York, p. 105, 1953.

OBSERVED RADIATION

E. Pettit: *The sun and stellar radiation.* In J. A. Hynek's: *Astrophysics.* McGraw-Hill Book Co., New York, Chapter 6, 1951.

HISTORY OF PHOTOMETRY

H. F. Weaver: *Pop. Astr.*, **54**, 211, 287, 339, 451, 504, 1946.

VISUAL PHOTOMETRY

J. Hellerich: *Visuelle Photometrie.* Handbuch der Experimental physik, Akademische Verlagsgesellschaft, Leipzig, **26**, 565, 1937.

PHOTOGRAPHIC PHOTOMETRY

K. Schwarzschild: *Publ. v. Kuffnerschen Sternw.*, **5**, 1899.

H. Kienle: *Photographische Photometrie.* Handbuch der Experimental physik, **26**, 649, 1937.

F. H. Seares: *P.A.S.P.*, **50**, 5, 1938.

E. W. H. Selwyn: *Photography in astronomy.* Eastman Kodak Co., 1950.

PHOTO-ELECTRIC PHOTOMETRY

Th. Walraven: *B.A.N.*, **11**, 421, 1952.

F. B. Wood: *Astronomical photo-electrical photometry.* A.A.A.S. Publ., 1953.

F. B. Wood: *The present and future of the telescope of moderate size.* Un. of Pennsylvania Press, Philadelphia, 1958.

LIGHT TIME

R. Prager: *Kleinere Veröff. Un. Sternwarte Berlin-Babelsberg* No. 12, 1932.

MAGNITUDE STANDARDS

F. H. Seares: *Trans. I.A.U.*, **1**, 69, 1922; See also **8**, 375, 1952.
H. L. Johnson and W. W. Morgan: *Ap.J.*, **117**, 313, 1953.
H. L. Johnson and D. L. Harris III: *Ap.J.*, **120**, 196, 1954.

PERIOD

E. Hertzsprung: *B.A.N.*, **4**, 178, 1928.
P. Th. Oosterhoff: *B.A.N.*, **5**, 37, 1929.
K. K. Kwee and H. van Woerden: *B.A.N.*, **12**, 327, 1956.

EXTINCTION

O. J. Eggen: *Ap.J.*, **111**, 68, 1950.
J. Stebbins, A. E. Whitford and H. L. Johnson: *Ap.J.*, **112**, 469, 1950.
S. Sharpless: *Ap.J.*, **116**, 254, 1952.
H. F. Weaver: *Ap.J.*, **116**, 638, 1952.

COLOR AND REDDENING

J. Stebbins, C. M. Huffer and A. E. Whitford: *Ap.J.*, **91**, 20, 1940; *Mt. W. Contr.* No. 621.
J. Stebbins and A. E. Whitford: *Ap.J.*, **98**, 20, 1943; **102**, 318, 1945.
A. E. Whitford: *Ap.J.*, **107**, 102, 1948.

POLARIZATION

J. S. Hall and A. H. Mikesell: *Publ. Naval Obs.*, **17**, part 1, 1950.
W. A. Hiltner: *Ap.J.*, **114**, 241, 1951; **120**, 454, 1954; *The Observatory*, **71**, 234, 1951.

VARIABLE STARS

K. Schiller: *Veränderlichen Sterne*. Verlag Barth, Leipzig, 1923.
P. W. Merrill: *The nature of variable stars*. The MacMillan Co., New York, 1938.
L. Campbell and L. Jacchia: *The story of variable stars*. Blakiston Co., Philadelphia, 1945.

OPEN CLUSTERS

R. J. Trumpler: *P.A.S.P.*, **37**, 307, 1925; *L.O.B.*, **14**, 154, 1930.
H. Shapley: *Star clusters*. McGraw-Hill Book Co., New York, 1930.
P. Collinder: *Ann. Obs. Lund*, No. 2, 1931.
H. L. Johnson and A. R. Sandage: *Ap.J.*, **121**, 616, 1955.
O. J. Eggen: *A.J.*, **60**, 401 and 407, 1955.

GLOBULAR CLUSTERS

P. ten Bruggencate: *Sternhaufen*, Springer Verlag, Berlin, 1927.
J. L. Greenstein: *Ap.J.*, **90**, 387, 1939.
H. B. Sawyer: *Publ. David Dunlap Obs.*, **1**, No. 20, 1947; **2**, No. 2, 1955.
H. C. Arp, W. A. Baum and A. R. Sandage: *A.J.*, **58**, 4, 1953.
A. R. Sandage: *A.J.*, **58**, 61, 1953.
W. A. Baum: *A.J.*, **59**, 422, 1954.

POPULATIONS

W. Baade: *Ap.J.*, **100**, 137, 1944; *Mt. W. Contr.*, No. 696, 1944.
D. J. K. O'Connell: *Ricerche Astronomiche*, **5**, 1958.

VI

Eclipsing Variables

97. *Units defined.* AN ECLIPSING VARIABLE HAS TWO components, which eclipse each other alternately. The method of determining the orbital elements from the light curve was initiated by H. N. Russell; only a few later refinements have been added. The extensive tables and the nomographs of J. E. Merrill facilitate the practical computation. The whole problem is very complex. We will start therefore with the simplest model and gradually make it more complicated, as stars really are. However, we will consider only the first approximation to keep the problem as simple as possible.

We have the same orbital elements as before, but now t_0 is the epoch of conjunction. From the light curve alone the longitude of the node Ω is indeterminable, just as in the case of a spectroscopic binary. This means that we can rotate the picture of the eclipsing system and still get the same light curve. We also cannot find the direction of the orbital motion from the light curve alone.

Moreover, if we imagine the geometry of the whole system to be enlarged by a factor of two while keeping the period constant, the light curve would remain the same. We will thus find relative sizes. If we should suddenly increase the intensities of both components by a factor of two, the shape of the light curve would remain the same. We find thus relative intensities and therefore we can take the sum of the intensities equal to unity. This can be seen in another way. The light of the variable is measured relative to the intensity of a comparison star. Naturally the orbital elements of the eclipsing system are completely independent of this comparison star chosen by the observer. Therefore we can measure the light of the variable relative to an imaginary comparison star of which the intensity equals the maximum light of the variable. We can thus shift the light curve by adding a constant in the magnitudes in order to make the magnitude difference in the maximum zero, and thus the intensity ratio equal unity. We define our units:

Length: Separation of the centers of the two components; the radius for a circular orbit, the semi-major axis for an elliptical orbit.

Area: The area of the small star.

Intensity: Sum of the intensities of both stars in the maximum of the light curve when they are uneclipsed.

In the following derivations we have to imagine that we can come so close to the stars that we can visualize the disks. There are at least 3000 known eclipsing variables, but only about 150 have good orbital determinations.

98. *Algol type.* We first consider the simplest case: circular orbit, spherical stars, uniform light, and no reflection effect. Because the orbit is circular, the minima are exactly $\frac{1}{2}P$ apart. Because the stars are spherical, the maximum has constant light. Uniform light over the disk means that the covered area is proportional to the loss of light. We will call:

r_s = radius of smaller star
r_g = radius of larger or greater star
L_s = intensity of smaller star
L_g = intensity of greater star
a = area covered, expressed in terms of the area of the small star as unit.

This last definition will have to be generalized later on. If we define k as the ratio of the radii and l_{max} as the maximum light we have:

$$k = \frac{r_s}{r_g} \leqslant 1, \qquad l_{max} = L_s + L_g = 1 \qquad (268)$$

The light we receive from the system is proportional to the sum of the brightness of the star in front and the brightness of the uncovered portion of the star behind. The decrease or loss of light is even more important because the shape of the observed minimum will tell us the relative sizes of the stars. We must use intensities throughout, and consequently we first convert the magnitude differences of the light curve into intensity ratios. Or if we take our above mentioned units, we have simply:

$$m = -2.5 \log l, \qquad \log l = -0.4\, m \qquad (269)$$

There are four possibilities of eclipses which occur in definite pairs. The names used are as follows:

TABLE II. POSSIBILITIES OF ECLIPSES

Eclipse	*a* *occultation*	*b* *transit*
1 Complete	total	annular
2 Incomplete	Partial	partial

The underlined letters are the abbreviations used. We shall consider the cases separately.

(1a) *total, occultation.* The greater star is in front, and during eclipse we can see only this one. The minimum of the light curve is therefore constant. The loss of light gives the brightness of the smaller star, and the intensity during minimum gives the brightness of the larger star (Figure 129):

$$1 - l_t = L_s, \qquad l_t = 1 - L_s = L_g \qquad (270)$$

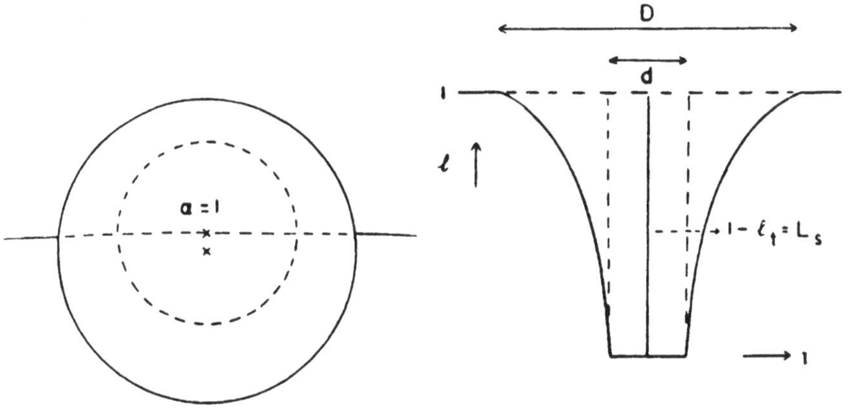

Figure 129. Total eclipse or occultation.

(1b) *annular, transit.* The smaller star is in front so that the loss of light is caused by a fraction of the light of the larger star being cut off. This fraction equals the ratio of the areas during mid-eclipse; thus $k^2 \leqslant 1$. Again the minimum has a constant or flat part. We find for the loss of light (Figure 130):

$$1 - l_a = k^2 L_g \tag{271}$$

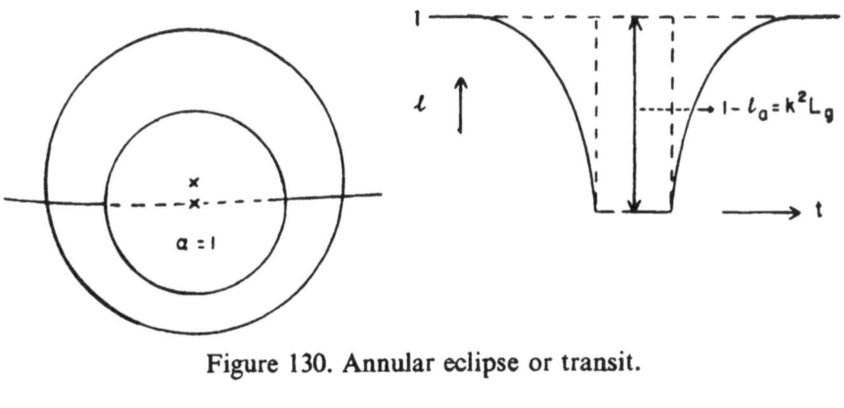

Figure 130. Annular eclipse or transit.

(2a) *Partial, occultation.* The greater star is in front. There is no longer constant light around mid-eclipse, but the minimum is curved; the loss of light at mid-eclipse is (Figure 131):

$$1 - l_P = a_0 L_s, \qquad a_0 = \frac{1 - l_P}{L_s} \tag{272}$$

This value of a_0 cannot be determined from the depth of only one minimum.

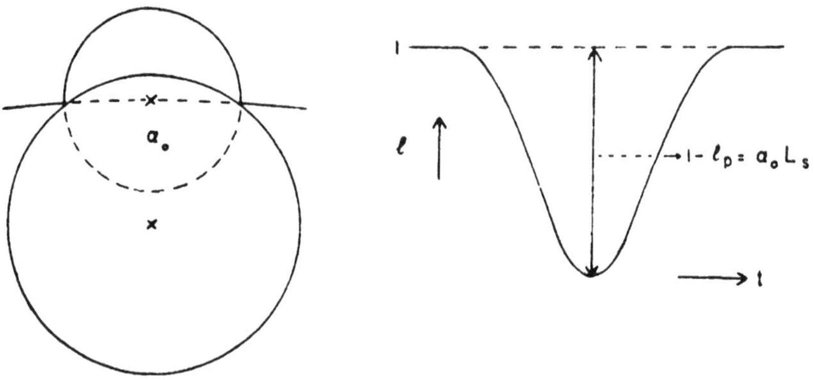

Figure 131. Partial eclipse with the greater star in front.

(2b) *partial, transit.* The smaller star is in front. Again the covered area or loss of light is expressed in terms of the smaller star as a unit. It equals $k^2\ a_0$ when expressed in terms of the greater star as a unit. This minimum is also curved; the loss of light at mid-eclipse is (Figure 132):

$$1 - l_p = k^2\, a_0 L_g \tag{273}$$

For both complete and incomplete eclipses we can proceed as follows: During mid-eclipses at the primary and the secondary minima the covered areas are the same, provided the orbit is circular. For uniform light we know that the loss of light in both cases is for areas of equal size. In other words, we are comparing surface luminosities in this model. This statement holds

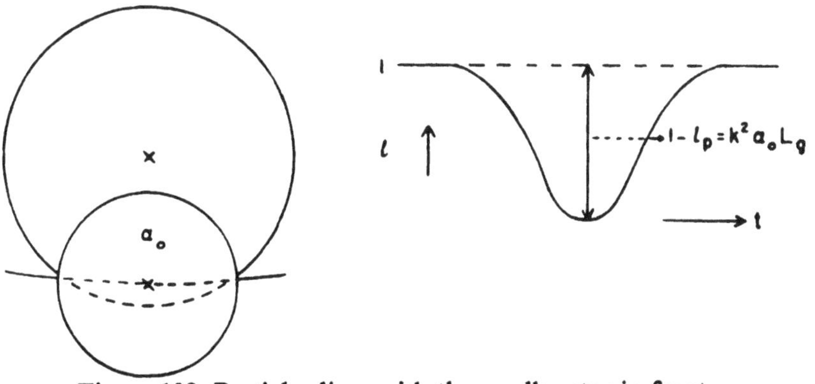

Figure 132. Partial eclipse with the smaller star in front.

also for points on the branch which have corresponding phases during the two minima, namely θ and 180° + θ. As cumulative effect we can consider the areas lying between the curved minima and the horizontal line of maximum in the light curve (areas minimum I and II). During primary minimum the loss of light is greater, and therefore this minimum corresponds to the eclipse of that star which has the higher surface luminosity and is thus the hotter star of the two.

$$\frac{I_h}{I_c} = \frac{1 - l_1}{1 - l_2} = \frac{\text{area minimum I}}{\text{area minimum II}} \tag{274}$$

99. *Loss of light.* Let us summarize what we have found for the loss of light during mid-eclipse and also find this loss of light for a point on the branch. See Figure 133.

TABLE III. LOSS OF LIGHT FOR UNIFORM CASE

Case	*mid-eclipse*	*branch*	*loss of light*
(1a) t	$1 - l_0 = L_s$	$1 - l = aL_s$	$a = \dfrac{1 - l}{1 - l_0}$
(1b) a	$= k^2 L_g$	$= k^2 a L_g$	
(2a) P	$= a_0 L_s$	$= aL_s$	$n = \dfrac{a}{a_0} = \dfrac{1 - l}{1 - l_0}$
(2b) p	$= k^2 a_0 L_g$	$= k^2 a L_g$	

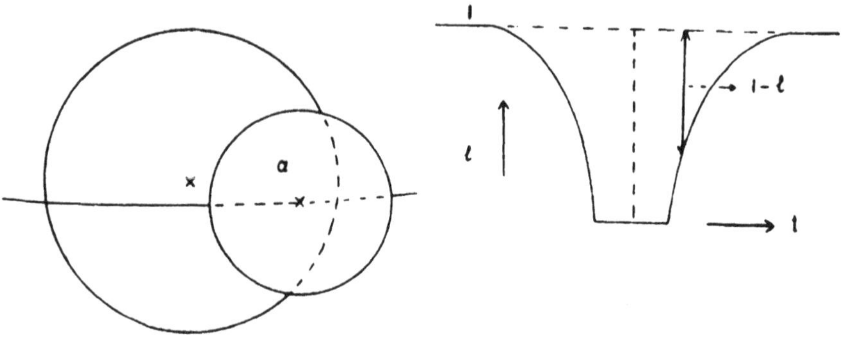

Figure 133. The loss of light.

For all four cases we can write the results in one formula:

$$n = \frac{\alpha}{\alpha_0} = \frac{1 - l}{1 - l_0} \tag{275}$$

Here $n = 1$ for mid-eclipse; $\alpha_0 = 1$ for complete eclipses, and $\alpha_0 \leqslant 1$ for partial eclipses. For partial eclipses at any normal point we can compute the loss of light or α, in terms of the maximum loss of light as unit.

100. *Depth relations.* We can get additional information from the depth of the minima if we use both of them.

(1) For complete eclipses we have:

$$(1 - l_t) + \frac{1 - l_a}{k^2} = L_s + L_g = 1$$

$$k^2 = \frac{1 - l_a}{l_t} \tag{276}$$

We allow only positive k so the formula in itself gives only one k. However, there are still two possibilities for the following reason. The annular eclipse can be either the primary minimum, which is the deeper, or the secondary minimum; the total eclipse is then the other one. Naturally for a given system only one possibility suffices. We can get some idea about this by considering the durations of the eclipses. If a system gives such a pair of eclipses, the inclination i will be close to 90°. If $i = 90°$, we have the longest durations of eclipses. Let the duration of time between the outer contacts be t_1 and between the inner contacts be t_2; then we have for each of the two minima:

$$\frac{t_1}{t_2} = \frac{2r_g + 2r_s}{2r_g - 2r_s}, \qquad \frac{t_1 - t_2}{t_1 + t_2} = \frac{4r_s}{4r_g} = k \tag{277}$$

The time interval between outer and inner contacts equals the time that the diameter of the smaller star needs to cross over or behind the edge of the larger star. These durations cannot be determined very accurately in practice, but they give us some idea about the value of k. Later on we will see how to find the correct k for the system by studying the shape of the branches. For only one of the two possible k's will we find the expected relation so that the problem for each individual system will be settled satisfactorily.

(2) For incomplete eclipses we have:

$$(1 - l_P) + \frac{(1 - l_p)}{k^2} = \alpha_0 (L_s + L_g) = \alpha_0$$

$$\alpha_0 = (1 - l_P) + \frac{(1 - l_p)}{k^2} = C + \frac{D}{k^2} \tag{278}$$

The loss of light at mid-eclipse varies linearly with $1/k^2$.

101. *The relation between* δ/r_g, k *and* α. During the eclipse we have $\delta < (r_g + r_s)$, when δ is the distance between the centers of the components projected on the plane of the sky. The covered area is just the same, independent of the particular star covered. We now find three equations:

$$\sin \varphi_1 = k \sin \varphi_2 \tag{279}$$

$$\delta = r_g(\cos \varphi_1 + k \cos \varphi_2) \tag{280}$$

$$k^2\pi\alpha = \varphi_1 + k^2\varphi_2 - k \sin (\varphi_1 + \varphi_2) \tag{281}$$

Starting with δ/r_g and k in the first two equations we find φ_1 and φ_2. Together with k these determine α from the third equation. Thus the above mentioned relation between the three quantities follows and can be tabulated. The proof of the third equation is (Figure 134):

$$\begin{aligned}\text{Area} &= \alpha\pi r_s^2 = \alpha\pi k^2 r_g^2\\ &= r_g^2(\varphi_1 - \cos\varphi_1 \sin\varphi_1) + r_s^2(\varphi_2 - \cos\varphi_2 \sin\varphi_2)\\ &= r_g^2(\varphi_1 - \cos\varphi_1 \sin\varphi_1 + k^2\varphi_2 - k^2 \cos\varphi_2 \sin\varphi_2)\\ &= r_g^2\{\varphi_1 + k^2\varphi_2 - k(\cos\varphi_1 \sin\varphi_2 + \cos\varphi_2 \sin\varphi_1)\}\\ &= r_g^2\{\varphi_1 + k^2\varphi_2 - k\sin(\varphi_1 + \varphi_2)\}\end{aligned}$$

Division by r_g^2 gives the equation desired.

The tables of these relations are given by:

E. Hetzer $\delta/r_g = f(k, \alpha)$, M. Wendt $\alpha = F(k, \delta/r_g)$

The latter author really gives $\alpha = F(k, \delta)$ where r_g is taken as a unit.

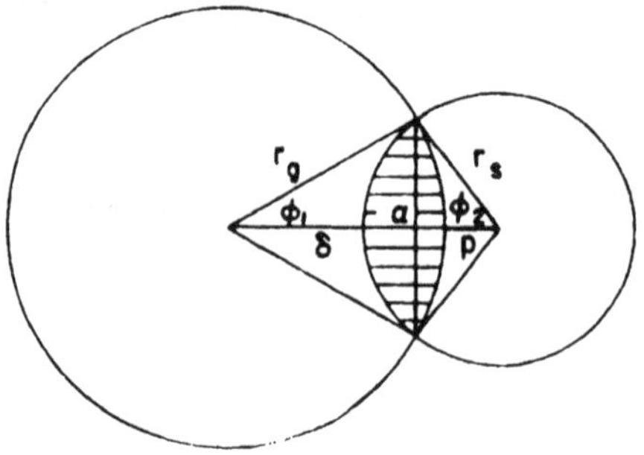

Figure 134. The relation between δ/r_g, $k = r_s/r_g$ and α.

Sometimes we use the geometrical depth to describe the eclipse. This geometrical depth p is defined as the shortest distance from the center of the smaller star to the edge of the greater star, and is expressed in terms of the radius of the smaller star as a unit.

$$p = \frac{\delta - r_g}{r_s}, \qquad \delta = r_g + pr_s = r_g(1 + kp) \qquad (282)$$

We have the following relation:

$$\frac{\delta}{r_g} = 1 + kp = f(k, \alpha) \tag{283}$$

102. *Dynamical condition.* If θ is the phase angle in the orbital plane from the middle of the minimum, and t_0 the time of mid-eclipse, we have:

$$\theta = \frac{2\pi}{P}(t - t_0), \qquad \theta = \text{phase} \times 360° \tag{284}$$

This is nothing else but the expression of the law of areas for a circular orbit. It can be computed for each normal point at time t. From the light curve or (m, t) curve we have converted into the (l, t) curve, but now we shall work with the (α, θ) curve. The last two coordinates can be calibrated on the graph of the intensity curve.

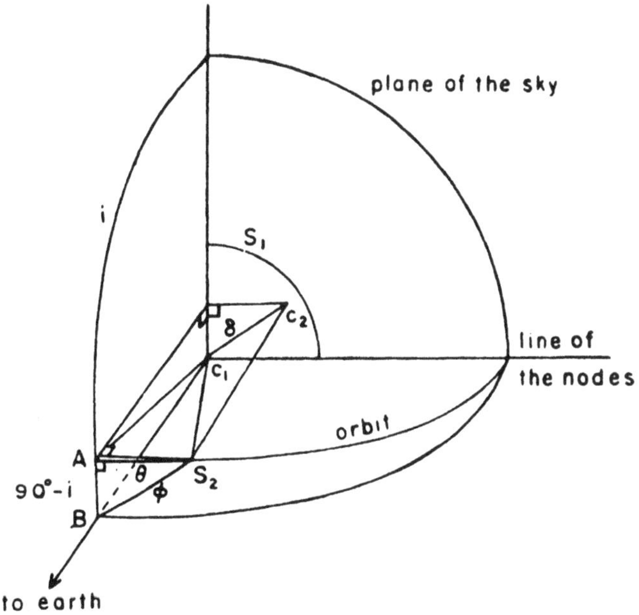

Figure 135. Derivation of the dynamical condition.

From the geometry of the eclipse we find from Figure 135:

$$\begin{aligned}\delta^2 &= \cos^2 i \cos^2\theta + \sin^2\theta = \cos^2\theta - \cos^2\theta \sin^2 i + \sin^2\theta \\ &= \cos^2\theta - \sin^2 i + \sin^2 i \sin^2\theta + \sin^2\theta \\ &= 1 - \sin^2 i + \sin^2 i \sin^2\theta\end{aligned}$$

$$\delta^2 = \cos^2 i + \sin^2 i \sin^2\theta \tag{285}$$

From this point on there are two methods of computation, which we shall distinguish as methods I and II. We shall now consider the shape of the branches.

103. *Determination of k, r_g, i with f function (method I).* This method is often followed in Europe. We write the dynamical condition in the following form:

$$\left(\frac{\delta}{r_g}\right)^2 = \frac{\cos^2 i}{r_g^2} + \frac{\sin^2 i}{r_g^2} \sin^2\theta = f^2(k, \alpha) \tag{286}$$

Let us call:

$$\left(\frac{\delta}{r_g}\right)^2 = u, \qquad \frac{\cos^2 i}{r_g^2} = A, \qquad \frac{\sin^2 i}{r_g^2} = B$$

The dynamical condition can then be written as:

$$u = A + B \sin^2\theta \tag{287}$$

There is thus a linear relation between u and $\sin^2\theta$ for the correct value of k and only for that value. We find this correct value of k in the following way (Figure 136).

Start with one of the two possible k's we have found from the depth relation. The α and θ are known for the normal points on the branch; the table gives the value of δ/r_g. Now plot $(\delta/r_g)^2 = u$ against $\sin^2\theta$ and see whether this is a linear relation. If we have chosen the wrong k from the two possibilities, the normal points will form a curved relation; we can disregard this k. Thus we know whether the annular eclipse corresponds to the primary or the secondary minimum of our particular system. To determine k better we make the same computation with some other values

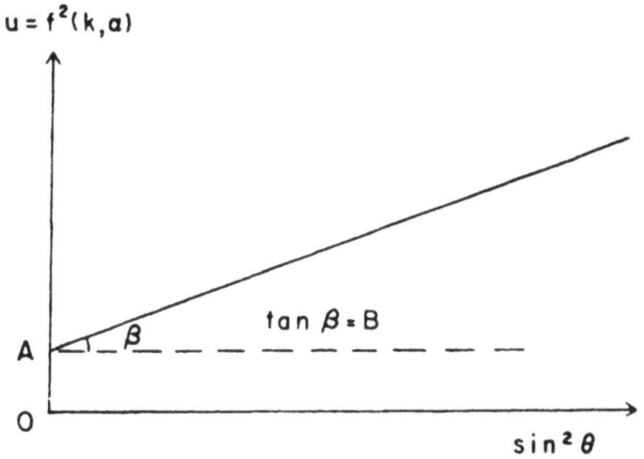

Figure 136. The linear relation according to method I.

of k in the neighborhood of the first one. After plotting the graph we see which one shows the best linear relation, and we decide that this will be the best possible value of k. We can obtain this somewhat better, since the linear relation should be computed by least squares. Here the weighting of the normal points is of importance. If the light curve is a photographic one, the magnitudes have the same weights on different points of the branch, but intensities do not. If it is a photo-electric curve, the measurements in intensity theoretically have the same weights, but the seeing effect works according to magnitude and disturbs the equal weights. Shifting the light curve somewhat in the magnitude direction changes α somewhat, and via the table we find the influence on δ/r_g and thus on u. The weight has to be taken inversely proportional to $(\Delta u)^2$.

The deviation of each normal point from the linear relation can be computed. It is even better to compute for each k the theoretical light curve in magnitudes and determine the residuals or $(O - C)$ for the normal points. It is clear that the best value for k occurs for a minimum value of $\Sigma(O - C)^2$. We plot this $\Sigma(O - C)^2$ against the corresponding k and bring a parabola with its axis parallel to the ordinate through the points (Figure 137). This can also be done by least squares. The minimum value of the parabola determines the best possible k.

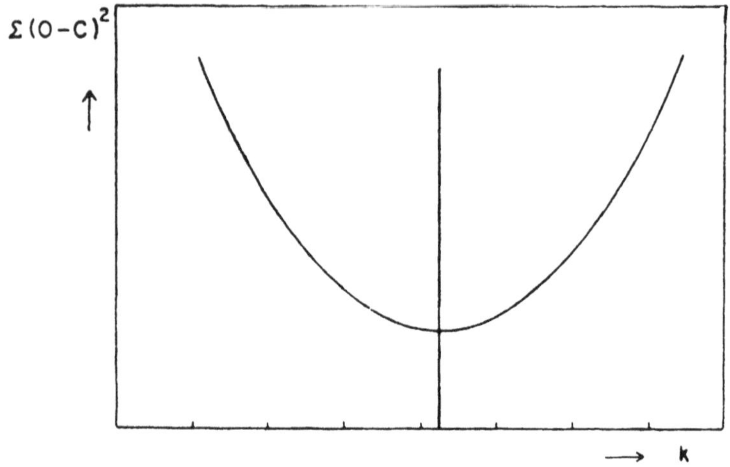

Figure 137. The minimum value of the parabola determines the best value of the ratio of the radii k.

From this best k and the given values of α and θ we again compute δ/r_g via the table. The relation $u = A + B \sin^2\theta$ is computed once more and gives a perfectly linear relation within the observational errors. From A and B we have:

$$\tan^2 i = \frac{B}{A}, \qquad \frac{1}{r_g^2} = A + B \qquad (288)$$

Thus i and r_g follow.

This method is quite impersonal, and two computers should get the same results. All the normal points are used and not the smooth curve through these points. For $i = 90°$ the dynamical condition becomes $\delta = \sin\theta$ for all k's. However, if a certain k gives $i = 90°$, then another k will show $i \neq 90°$ in order to give the same relation. To determine k the first dynamical condition in its general form must be used in all cases.

For partial eclipses we have to work with $n = \alpha/\alpha_0$ and θ. In some cases we find k in the same way and then α_0 from the depth relation. In other cases we can find k only as a function of n or α_0. This we shall call the shape relation. Combined with the depth relation we can solve for k and α_0 separately. With these two quantities known, we find L_s from: $1 - l_P = \alpha_0 L_s$, and L_g from: $L_s + L_g = 1$.

104. *Differential corrections.* The magnitude of a system at a given phase depends on the orbital elements k, r_g, i or on A, B, k. We can then write:

$$m(A+dA, B+dB, k+dk) = m(A, B, k) + \frac{\partial m}{\partial A}dA + \frac{\partial m}{\partial B}dB + \frac{\partial m}{\partial k}dk$$

(A, B, k) is the set of preliminary elements that gives the differences $(O - C)$ between the observed and computed normal points of the light curve expressed in magnitudes. Then:

$$O - C = \frac{\partial m}{\partial A}dA + \frac{\partial m}{\partial B}dB + \frac{\partial m}{\partial k}dk \tag{289}$$

Each normal point gives an equation of this form. The corrections dA, dB, dk, to the provisional elements are found by a least squares solution.

We can find the coefficients with help of the dynamical condition:

$$u(k, \alpha) = A + B\sin^2\theta = f^2(k, \alpha)$$

Successively holding constant two of the three variables we get:

$$\left.\begin{aligned} &u(k \quad\;\; , \alpha + \Delta\alpha) - u(k, \alpha) = \Delta A \\ &u(k \quad\;\; , \alpha + \Delta\alpha) - u(k, \alpha) = \Delta B \sin^2\theta \\ &u(k+\Delta k, \alpha + \Delta\alpha) - u(k, \alpha) = 0 \end{aligned}\right\}$$

The last equation can be written as: $f(k+\Delta k, \alpha + \Delta\alpha) = f(k, \alpha)$

$$\left.\begin{array}{ll} \Delta u(\alpha) = \Delta A = \Delta B \sin^2\theta & \\ \Delta u(k, \alpha) = 0, & \Delta f(k, \alpha) = 0 \end{array}\right\} \tag{290}$$

With help of the table we can find for a given $\Delta\alpha$ the corresponding $\Delta A = \Delta B \sin^2\theta$ and Δk and thus computed values of:

$$\frac{\partial\alpha}{\partial A}, \qquad \frac{\partial\alpha}{\partial B} = \frac{\partial\alpha}{\partial A}\sin^2\theta, \qquad \frac{\partial\alpha}{\partial k}$$

We have still to find an expression for $\partial m/\partial\alpha$.

$$m = -2.5 \log l = -2.5 \operatorname{Mod} \ln l = -1.0857 \ln l$$

$$\frac{\partial m}{\partial l} = -\frac{1.0857}{l} \tag{291}$$

When the intensity at mid-eclipse is l_0, we have for complete eclipses:

$$\alpha = \frac{1-l}{1-l_0}, \qquad 1 - l = \alpha(1 - l_0), \qquad -\frac{\partial l}{\partial\alpha} = 1 - l_0$$

$$\frac{\partial m}{\partial\alpha} = \frac{\partial m}{\partial l}\frac{\partial l}{\partial\alpha} = 1.0857\frac{(1-l_0)}{l} \tag{292}$$

By multiplication we now find: $\partial m/\partial A$, $\partial m/\partial B$, $\partial m/\partial k$. It is also possible to introduce corrections for the intensity and limb darkening. The differential corrections have been given by L. Binnendijk, A. B. Wyse, and G. E. Kron.

105. *Complete eclipses with ψ and φ functions (method II).* This method is the original method of H. N. Russell and is still often used in the United States. The dynamical condition is now written in the form:

$$\delta^2 = \cos^2 i + \sin^2 i \sin^2\theta = r_g^2(1+kp)^2 \tag{293}$$

We bring a smooth curve through the normal points on the branch and will continue to use this curve (Figure 138). We choose a point on the curve with the coordinates $\alpha_2 = 0.6$, θ_2 and another point with coordinates $\alpha_3 = 0.9$, θ_3. For the given

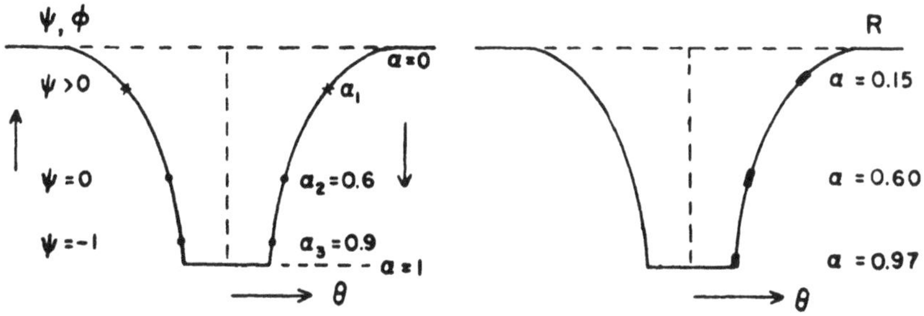

Figure 138. The standard points of the minimum of a complete eclipse.

system we can read off the values θ_2 and θ_3. These we will call our standard points. Let α_1, θ_1 denote a running point on the curve. We have then:

$$\sin^2 i\,(\sin^2\theta_1 - \sin^2\theta_2) = r_g^2\{(1+kp_1)^2 - (1+kp_2)^2\}$$
$$\sin^2 i\,(\sin^2\theta_3 - \sin^2\theta_2) = r_g^2\{(1+kp_3)^2 - (1+kp_2)^2\}$$

By division:

$$\frac{\sin^2\theta_1 - \sin^2\theta_2}{\sin^2\theta_2 - \sin^2\theta_3} = \frac{(1+kp_1)^2 - (1+kp_2)^2}{(1+kp_2)^2 - (1+kp_3)^2}$$

Our two standard points also determine the geometrical depths p_2 and p_3. Further, the expression $(1+kp)$ is a function of k and α, as we saw before.

If we take $\theta_1 = \theta$ for our running point, we can write:

$$\frac{\sin^2\theta - \sin^2\theta_2}{\sin^2\theta_2 - \sin^2\theta_3} = \psi\,(k, \alpha) \qquad (294)$$

This function was first given by H. N. Russell. For the standard point (α_2, θ_2) it follows that $\psi(k, \alpha_2) = 0$. For the standard point (α_3, θ_3) we have $\psi(k, \alpha_3) = -1$. From our two standard points we can find two positive constants:

$$\sin^2\theta_2 = A', \qquad \sin^2\theta_2 - \sin^2\theta_3 = B' \qquad (295)$$

The primes are added here to show that these constants are quite different from those found in the first method. Thus:

$$\frac{\sin^2\theta - A'}{B'} = \psi(k, \alpha) \qquad \text{For } \theta = 0, \quad \psi(k, \alpha_0) = -\frac{A'}{B'}$$

$$\sin^2\theta = A' + B'\,\psi(k, \alpha) \tag{296}$$

This is again a linear relation, but here A' and B' are known (Figure 139). For a certain point on the curve α and θ are known, so that $\psi(k, \alpha)$ follows, and via the table we find the particular k that characterizes the theoretical light curve going exactly through this point and the two standard points. It is clear that

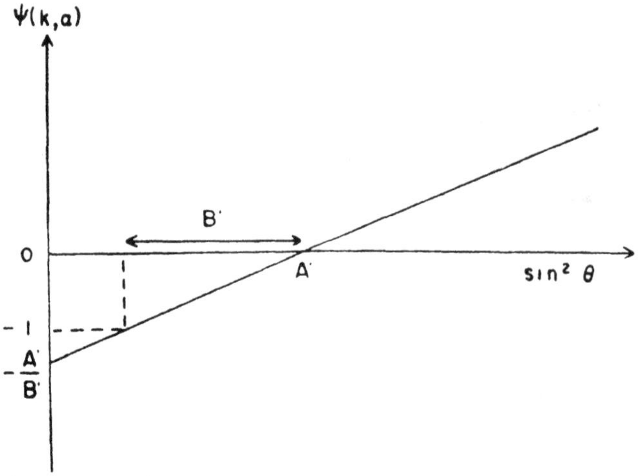

Figure 139. The linear relation according to method II.

if the third running point is taken too close to one of the standard points, k must be given small weight, and the corresponding theoretical light curve may deviate considerably from the observed one. The best procedure is to consider a number of points in the interval $\alpha = 0.4$ to $\alpha = 0$, where ψ is positive, each determining a value of k. The weighted average value is then taken as the best value of k. With this k and some values in its close neighborhood, a number of theoretical light curves are computed to see which k gives the best fit in magnitudes. In this way the criticism is overcome that the choice of the two original standard points may be affected by drawing errors.

A still better procedure is the following. We take three groups of readings and give them the appropriate weights in parentheses.

$\alpha = 0.05\,(2),\ 0.10\,(2),\ 0.20\,(2),\ 0.30\,(1),\quad \text{mean } \alpha = 0.14$

$\alpha = 0.50\,(1),\ 0.60\,(1),\ 0.70\,(1),\quad \text{mean } \alpha = 0.60$

$\alpha = 0.95\,(1),\ 0.97\,(2),\ 0.985\,(2),\quad \text{mean } \alpha = 0.97$

The theoretical light curve going through these three points or mean values naturally gives the same kind of formulae, but another function $R(k)$ from which the best k follows immediately. This function has been computed by J. E. Merrill.

We have still to find the elements r_g and i. For external contact we have $\delta = r_g + r_s$ and $p = +1$; for internal contact we have $\delta = r_g - r_s$ and $p = -1$. If we consider these points respectively, we find:

$$\left.\begin{aligned} \cos^2 i + \sin^2 i \sin^2\theta_e &= r_g^2(1+k)^2 \\ \cos^2 i + \sin^2 i \sin^2\theta_i &= r_g^2(1-k)^2 \end{aligned}\right\} \tag{297}$$

These two equations are solved as follows. Subtraction gives:

$$\sin^2 i\,(\sin^2\theta_e - \sin^2\theta_i) = r_g^2\, 4k$$

$$r_g^2 = \frac{1}{4k}\sin^2 i\,(\sin^2\theta_e - \sin^2\theta_i) \tag{298}$$

Substituting this value of r_g^2 in the first dynamical condition gives:

$$\cos^2 i + \sin^2 i \sin^2\theta_e = \frac{1}{4k}\sin^2 i\,(\sin^2\theta_e - \sin^2\theta_i)\,(1+k)^2$$

$$\cot^2 i + \sin^2\theta_e = \frac{1}{4k}(\sin^2\theta_e - \sin^2\theta_i)(1+k)^2$$

$$\cot^2 i = \frac{1}{4k}\left\{(1-k)^2\sin^2\theta_e - (1+k)^2\sin^2\theta_i\right\} \tag{299}$$

We now compute and thus do not read off from the curve:

$$\sin^2\theta_e = A' + B'\psi(k, 0), \qquad \sin^2\theta_i = A' + B'\psi(k, 1) \tag{300}$$

We substitute these in the formulae:

$$\frac{r_g^2}{\sin^2 i} = \frac{1}{4k}\{A' + B'\psi(k, 0) - A' - B'\psi(k, 1)\}$$

$$= \frac{B'}{4k}\{\psi(k, 0) - \psi(k, 1)\}$$

$$\cot^2 i = \frac{1}{4k}[(1-k)^2\{A' + B'\psi(k, 0)\} - (1+k)^2\{A' + B'\psi(k, 1)\}]$$

$$= \frac{B'}{4k}\{(1-k)^2\psi(k, 0) - (1+k)^2\psi(k, 1)\} - A'$$

We can write these results in the following form:

$$\frac{r_g^2}{\sin^2 i} = \frac{B'}{\varphi_1(k)}, \qquad \cot^2 i = \frac{B'}{\varphi_2(k)} - A' \tag{301}$$

The functions φ_1 and φ_2 depend only on k and are tabulated. With their help we can easily find i and r_g in practice.

In this method we measured two standard points and one suitable running point and found k, i, r_g. This k having been found from the branch rules out one of the two possible values found from the depth relation.

106. *Incomplete eclipses with χ and q functions (method II).* For a partial eclipse we cannot work with α but have to introduce n. Our previous formula (296) becomes:

$$\sin^2\theta(n) = A' + B'\psi(k, n, \alpha_0)$$

In addition to k, i, r_g we have to determine the element α_0. For a shallow minimum this is nearly impossible if photometric observations alone are available. For a deep minimum, we can determine the relation between k and α_0 from the branch; this we will call the shape relation. We can combine this with the depth relation between k and α_0. The intersection of both relations gives k and α_0 separately. Sometimes even this does not give a clear solution in practice. The best procedure then is to start with a combination k, α_0 according to the depth relation, and compute a set of theoretical light curves to compare with the observed one.

It is found practical to adopt two different standard points, namely those at mid-eclipse $n = 1$, $\theta = 0$ and at the half depth of the minimum $n = \frac{1}{2}$, $\theta(\frac{1}{2})$. This last value can be read off from

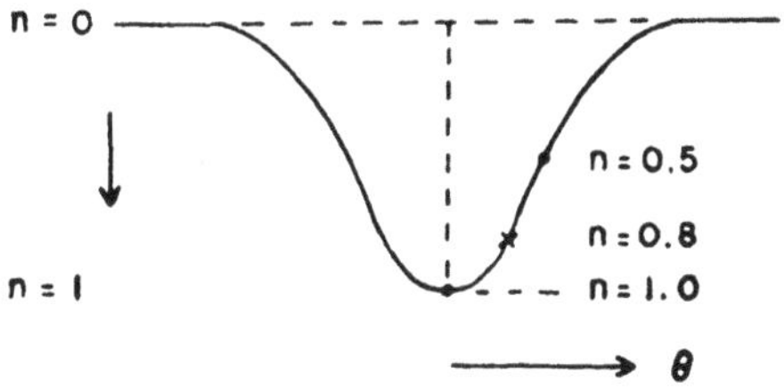

Figure 140. The standard points of the minimum of a partial eclipse.

the smooth curve of the minimum. Naturally we now get another function, χ (k, n, α_0). Again we take the third point n, $\theta(n)$ as our running point. The dynamical condition becomes (Figure 140):

$$\cos^2 i + \sin^2 i \sin^2\theta(n) = r_g^2 (1 + kp_n)^2$$
$$\cos^2 i = r_g^2 (1 + kp_0)^2 \quad \text{mid-eclipse}$$
$$\cos^2 i + \sin^2 i \sin^2\theta(\tfrac{1}{2}) = r_g^2 (1 + kp_{\frac{1}{2}})^2 \quad \text{half depth}$$

Subtraction gives:

$$\sin^2 i \sin^2\theta(n) = r_g^2 \{(1 + kp_n)^2 - (1 + kp_0)^2\}$$
$$\sin^2 i \sin^2\theta(\tfrac{1}{2}) = r_g^2 \{(1 + kp_{\frac{1}{2}})^2 - (1 + kp_0)^2\}$$

Division gives:

$$\frac{\sin^2\theta(n)}{\sin^2\theta(\frac{1}{2})} = \frac{(1 + kp_n)^2 - (1 + kp_0)^2}{(1 + kp_{\frac{1}{2}})^2 - (1 + kp_0)^2}$$

Here again the geometrical depth is known as a function of k and α. The relation among k, α, p has been computed by J. E. Merrill, who tabulated the right-hand side of the equation:

$$\frac{\sin^2\theta(n)}{\sin^2\theta(\frac{1}{2})} = \chi\,(k, n, \alpha_0) \tag{302}$$

This function is always positive.

$$\sin^2\theta(n) = \sin^2\theta(\tfrac{1}{2}).\,\chi\,(k, n, \alpha_0) \tag{303}$$

The form of this formula is to be expected because the choice of our standard points determines that $\sin^2\theta_2 = A' = 0$, and also $B' = \sin^2\theta(\frac{1}{2})$.

In the table for χ we enter with a value of n, corresponding to a point on the curve, not too close to the standard points. We thus find the shape relation between k and α_0. The depth relation is here a special case of the q functions which hold for limb darkening. Combination of the two relations gives k and α_0. We find L_s from the formula: $1 - l_P = \alpha_0 L_s$ and L_g from: $L_s + L_g = 1$. Some partial eclipses are so shallow that we cannot find definite elements.

For a better determination we can take the mean value of $\sin\theta$ for $n = 0.4, 0.5, 0.6$ as one of our standard points, and the other one again at mid-eclipse. Take as a third point $n = 0.8$. The theoretical light curve will go through $n = 0.5, 0.8$, and 1.0. Our function becomes:

$$\chi(k, \alpha_0) = \frac{\sin^2\theta(0.8)}{\sin^2\theta(0.5)} \tag{304}$$

It is clear that this can also be done with complete eclipses, at least in most cases. The nomograph of J. E. Merrill is drawn in this way and furnishes us a quick solution.

For external contact and mid-eclipse we have respectively:

$$\cos^2 i + \sin^2 i \sin^2\theta_e = r_g^2\,(1 + k)^2 \qquad \text{external}$$
$$\cos^2 i = r_g^2\,(1 + kp_0)^2 \qquad \text{mid-eclipse}$$

Here the θ_e, k, p_0 are known quantities. By subtraction we get:

$$\sin^2 i \sin^2\theta_e = r_g^2 (2k + k^2 - 2kp_0 - k^2 p_0^2)$$
$$= r_g^2 k (2 + k - 2p_0 - kp_0^2)$$

$$\sin^2 i = \frac{r_g^2 k}{\sin^2\theta_e} (2 + k - 2p_0 - kp_0^2) = r_g^2 u'$$

Again the relation among k, a_0, p_0 is given. Combination with the formula for mid-eclipse gives:

$$1 = r_g^2 (1 + kp_0)^2 + r_g^2 u' = r_g^2 (1 + 2kp_0 + k^2 p_0^2 + u') = r_g^2 v'$$

$$r_g^2 = \frac{1}{v'}$$

The u' and v' are known quantities; we thus find r_g as a positive quantity; substitution in the expression for mid-eclipse gives i.

Tables facilitating the differential corrections of the Russell method have been given by J. B. Irwin.

TABLE IV. SUMMARY FOR UNIFORM LIGHT

Eclipse	*occultation*	*transit*	*depth*	*branch*
complete	total	annular	I_h/I_c	
$x = 0$	flat	flat	$k^2 = \frac{1 - l_a}{l_t}$	
	α, θ	α, θ	L_s, L_g	
method I	f, A, B	f, A, B		k, r_g, i
,, II	ψ, φ, (R)	ψ, φ, (R)		
	χ	χ		
incomplete	Partial	partial	I_h/I_c	
$x = 0$	curved	curved	$a_0 = C + \frac{D}{k^2}$	
	n, θ	n, θ		
method I	f, A, B	f, A, B		$a_0 \longleftrightarrow k$
,, II	χ	χ	a_0, k, r_g, i, L_s, L_g	

107. *Limb darkening.* We consider the case: circular orbit, spherical stars, and limb darkening. A photograph of the sun shows the limb darker than the center of the disk; a color picture or visual comparison shows the disk redder at the limb. In the center of the solar disk we can look into the sun a certain distance, corresponding to a layer with a certain temperature. Near the edge of the disk we look into the sun a comparable distance but the deepest layer we can see there is a higher layer, and therefore is cooler and redder. The limb darkening is thus a function of the wavelength of light as well as the temperature of the star, $x = f(\lambda, T)$. A higher degree of limb darkening is found for the late spectral type stars than for the *B* and *A* type stars (Figure 141).

From the theory of radiation transfer in stellar atmospheres the following formula can be derived for limb darkening. This agrees very well with the observations of the sun for which

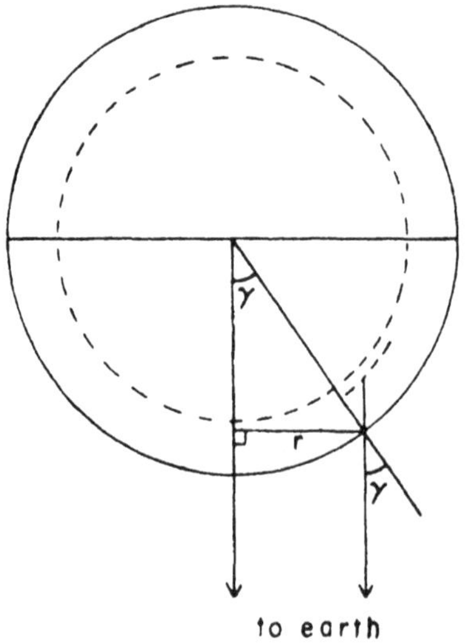

Figure 141. Explanation of the limb darkening effect of the sun.

$x = 0.6$ for photographic light. The I is the surface luminosity, and γ the angle between the radius and the line of sight.

$$I_x = I_0(1 - x + x\cos\gamma) \qquad (305)$$

For the coefficient of limb darkening we have $0 \leqslant x \leqslant 1$.
For uniform light: $x = 0$, $I = I_0$
For total limb darkening: $x = 1$, $I_1 = I_0 \cos\gamma$
We can therefore rewrite the formula for limb darkening in the form of a light balance. (Figure 142):

$$I_x = (1 - x)\, I_0 + x\, I_1 \qquad (306)$$

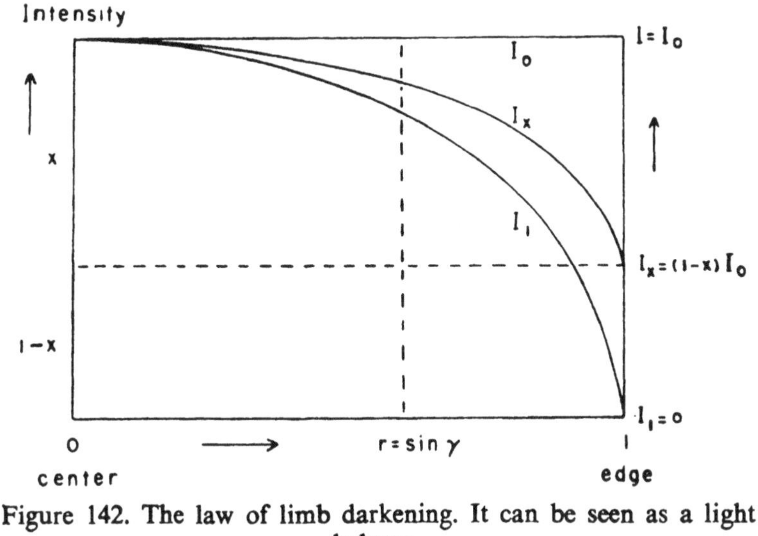

Figure 142. The law of limb darkening. It can be seen as a light balance.

If the surface brightness is known for uniform light and at any point of the star for total darkening at the limb, it can be found for intermediate darkening. In this formula the units are naturally the same for $x = 0$ and $x = 1$.

Let us consider the effect of total limb darkening on a spherical star whose total uniform light is $L_0 = \pi I_0$. In a circular ring of radius $r = \sin\gamma$ the surface brightness is the same because of symmetry (Figure 143). The area is $2\pi r\,dr = 2\pi \sin\gamma \cos\gamma\, d\gamma$. For the total light in the ring we have to multiply by $I_0 \cos\gamma$, thus:

$$2\pi r I_0 \cos\gamma\, dr = 2\pi I_0 \sin\gamma \cos^2\gamma\, d\gamma$$

The total light within a circle of radius r is:

$$2\pi \int_0^r r I_0 \cos\gamma\, dr = 2\pi I_0 \int_1^{\cos\gamma} \cos^2\gamma \sin\gamma\, d\gamma$$

$$= -2\pi I_0 \int_1^{\cos\gamma} \cos^2\gamma\, d\cos\gamma = -2\pi I_0 \tfrac{1}{3} \cos^3\gamma \Big/_1^{\cos\gamma}$$

$$= \pi I_0 \tfrac{2}{3}(1 - \cos^3\gamma) = L_0 \tfrac{2}{3}(1 - \cos^3\gamma)$$

Over the whole disk we have to integrate from $r = 0$ to 1, or from $\cos\gamma = 1$ to 0. Thus $L_1 = \frac{2}{3} L_0$. For total limb darkening the star has only $\frac{2}{3}$ the intensity of uniform light.

For intermediate limb darkening we find:

$$L_x = (1 - x)L_0 + xL_1 = (1 - x)L_0 + x\tfrac{2}{3}L_0 = (1 - \tfrac{1}{3}x)L_0 \quad (307)$$

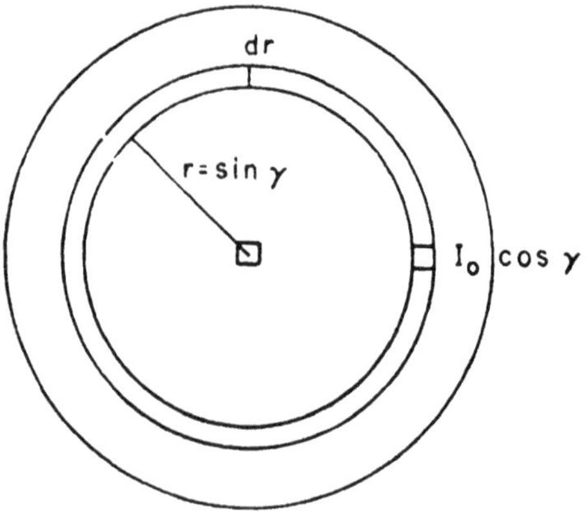

Figure 143. For total limb darkening a spherical star has only 2/3 the intensity of uniform light.

Thus for intermediate limb darkening the star has only the fraction $(1 - \frac{1}{3}x)$ of its uniform light. Consequently it lost the fraction $\frac{1}{3}x$ of its uniform light.

108. *Influence of limb darkening.* We must consider the four cases of minima. In every case the effect of limb darkening is to diminish the total depth during mid-eclipse. From a practical standpoint this is not interesting, since we have only observed minima of a given depth. During an annular eclipse, however, we do have a change. The minimum itself is now curved because not always the same amount of light of the larger star is blocked off during the transit. Figure 144 shows that we need another model of different size, and thus different orbital elements, to

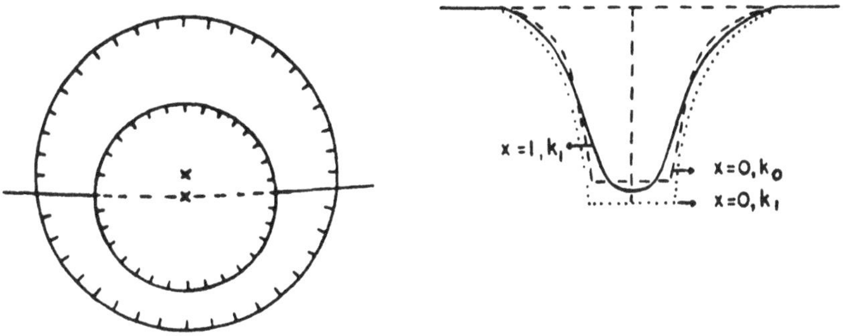

Figure 144. The limb darkening causes the minimum of an annular eclipse to be curved. We need another model of different size and thus different orbital elements than in the case of uniform light.

give a theoretical light curve in agreement with the observations. The minimum of total eclipse remains flat so that a combination of a curved and a flat minimum proves limb darkening of the larger star. For the partial eclipses there are changes around mid-eclipse, but these minima were already curved. The ratio of the surface brightnesses cannot be found from the ratio of the depths, since the loss of light is no longer proportional to the covered area. The mean ratio of surface brightnesses can be computed after the elements have been derived.

We now have to define α in a different way than for uniform light. A logical way would be to define α as the loss of light, expressed in terms of the maximum loss of light at mid-eclipse as a unit. For a total eclipse and for partial eclipses this definition holds. However, for tabular purposes in the case of an annular eclipse we define α as follows:

α = loss of light, expressed in terms of the loss of light at internal tangency as a unit (Figure 145).

For an annular eclipse we introduce:

$$\alpha = \frac{1 - l}{1 - l_i}, \qquad \tau_a = \frac{1 - l_i}{L_g}, \qquad \tau_a \alpha = \frac{1 - l}{L_g} \tag{308}$$

For annular mid-eclipse one gets: $1 - l_0 = \tau_a \alpha_0 L_g = q_0 L_g$. The $\tau_a \alpha$ is the loss of light expressed in terms of the intensity of the larger star as a unit. For the total eclipse there is no need to introduce τ because:

$$\tau_t = \frac{1 - l_t}{L_s} = 1, \qquad \text{since } l_t = L_g.$$

In general the loss of light for occultation and transit is not the same, and we must introduce α^{oc} and α^{tr}. Both quantities

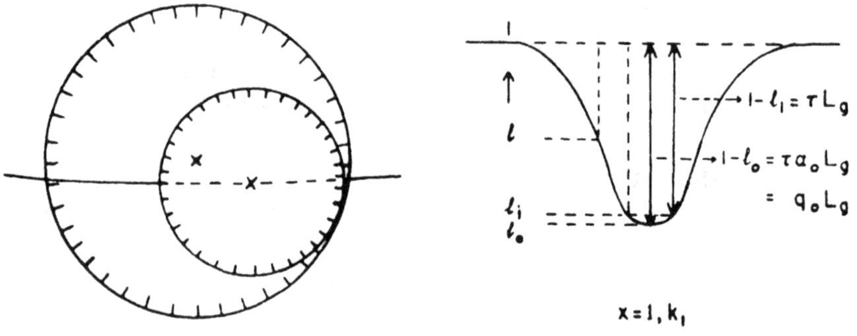

Figure 145. The unit of loss of light is taken at internal tangency.

are zero for external tangency ($p = +1$), and both are unity for internal tangency ($p = -1$). See Figure 146. For the relation among δ/r_g, k, α or p, k, α we divide the area into rings of equal brightness and add the brightnesses of all the covered rings to find the loss of light. This is a question of numerical integration. The same technique holds for partial eclipses.

Tables for total limb darkening have been given by H. N. Russell and H. Shapley, W. P. Zessewitsch, K. Ferrari. Extensive tables for intermediate limb darkening ($x = 0.0, 0.2, 0.4, 0.6, 0.8, 1.0$) have been given for method I by W. P. Zessewitsch and for method II by J. E. Merrill.

TABLE V. LOSS OF LIGHT FOR LIMB DARKENING

Case	*mid-eclipse*	*internal tangency*	*branch*	*loss of light*
(1a) t	$1-l_0= \; 1\,L_s$	$1-l_i= \; L_s$	$1-l= \alpha^{oc}L_s$	$\alpha^{oc}=\dfrac{1-l}{1-l_i}$
(1b) a	$=\tau\alpha_0^{tr}\,L_g$	$=\tau L_g$	$=\tau\alpha^{tr}L_g$	$\alpha^{tr}=\dfrac{1-l}{1-l_i}$
(2a) P	$=\alpha_0^{oc}\,L_s$	—	$=\alpha^{oc}L_s$	$n^{oc}=\dfrac{\alpha^{oc}}{\alpha_0^{oc}}=\dfrac{1-l}{1-l_0}$
(2b) p	$=\tau\alpha_0^{tr}\,L_g$	—	$=\tau\alpha^{tr}L_g$	$n^{tr}=\dfrac{\alpha^{tr}}{\alpha_0^{tr}}=\dfrac{1-l}{1-l_0}$

To describe the minimum of the annular eclipse around the time of mid-eclipse we define in general:

$$q = \tau\frac{\alpha^{tr}}{\alpha^{oc}} = \frac{1-l^{tr}}{L_g} : \frac{1-l^{oc}}{L_s} \tag{309}$$

At mid-eclipse we have $\alpha_0^{oc} = 1$ and $q_0 = \tau\alpha_0^{tr}$. This last relation is thus a special case of a more general definition. This q_0 is important in the depth relation.

For complete eclipses we get for mid-eclipses:

(1a) $1 - l_0 = 1 - l_i = 1 - l_t = L_s\,,$ $\qquad \alpha_0^{oc} = 1$

(1b) $1 - l_0 = 1 - l_a = \tau\alpha_0^{tr}\,L_g = q_0\,L_g\,,$ $\qquad \alpha_0^{tr} > 1$

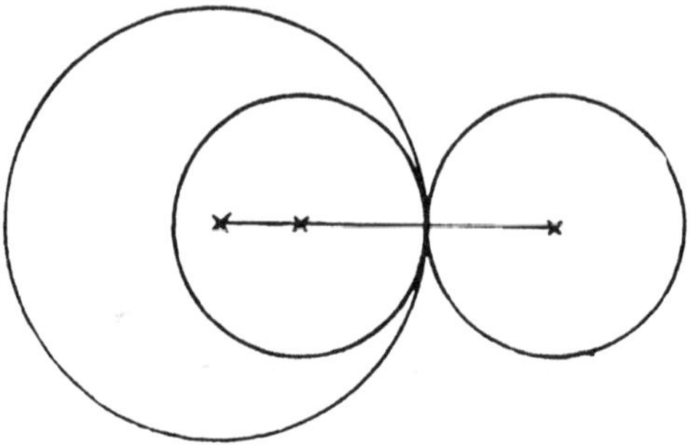

Figure 146. Loss of light and geometrical depth at internal and external tangency.

The depth relation becomes:

$$1 - l_t + \frac{1 - l_a}{q_0} = L_s + L_g = 1$$

$$q_0 = \frac{1 - l_a}{l_t} \tag{310}$$

For partial eclipses we get:

(2a) $1 - l_0 = 1 - l_P = \alpha_0^{oc} L_s$

(2b) $1 - l_0 = 1 - l_p = \tau \alpha_0^{tr} L_g = q_0 \alpha_0^{oc} L_g$

For partial eclipses the depth relation becomes:

$$1 - l_P + \frac{1 - l_p}{q_0} = \alpha_0^{oc} (L_s + L_g) = \alpha_0^{oc}$$

$$\alpha_0^{oc} = (1 - l_P) + \frac{(1 - l_p)}{q_0} = C + \frac{D}{q_0} \tag{311}$$

For uniform light we have: $\alpha^{tr} = \alpha^{oc} = \alpha$, and $q_0 = \tau = k^2$, which gives the old formulae. For example, for an annular eclipse $\tau = (1 - l_i)/L_g$ gives the previous formula $k^2 = (1 - l_a)/L_g$.

The depth relation gives a relation between α_0^{oc} and q_0, the latter quantity being related to k^2. Via a table we find the relation between α_0^{oc} and k. For complete eclipses $\alpha_0^{oc} = 1$, and the formula transforms into the form of the above mentioned one.

The dynamical condition remains the same. However, we must use the new functions, which are different from the old ones for uniform light, and we must remember that they differ for occultation and transit. For complete eclipses one can determine the best x and so find the limb darkening. For partial eclipses, a value of x is assumed based upon the spectral types if these are known. We find the shape relation between α_0 and k; the combination with the depth relation gives α_0 and k separately. Sometimes this does not give a clear solution. We start then with a combination q_0, α_0 or k, α_0 according to the depth relation, and compute a set of theoretical light curves to compare with the observed one.

Because the temperatures of the components are usually different, they will have different colors, so that even with uniform light the combined color changes during eclipse. During primary minimum the hottest star is eclipsed and the coolest one is in front. During this minimum the combined color of the system is becoming redder in general. This can be computed in detail by giving the single colors weights proportional to the momentary light. However, the different temperatures cause different limb darkening for the components. They can be found by analyzing the branches of each minimum separately. In case of limb darkening the components are redder near the edge and this gives an additional change in the combined color during eclipse.

The limb darkening of one component is different in photographic and photovisual light. From the photographic light curve of a minimum we can determine the x_{pg} for the star; from the photovisual light curve the x_{pv} for the same component.

This means therefore that the combined color curve determines the $x_{pg} - x_{pv}$ of the component.

TABLE VI. SUMMARY FOR LIMB DARKENING

Eclipse	*occultation*	*transit*	*depth*	*branch*
complete	total	annular		
$x=1$	flat	curved	$q_0=\frac{1-l_s}{l_t}$	
	α, θ	α, τ, θ	L_s, L_g	
I	f^{oc}, A, B	f^{tr}, A, B		x, k, r_g, i
$x \leqslant 1$; II	ψ^{oc}, ϕ^{oc}	ψ^{tr}, ϕ^{tr}		
	χ^{oc}	χ^{tr}		
incomplete	Partial	partial		
$x=1$	curved	curved	$\alpha_0=C+\frac{D}{q_0}$	
	n, θ	n, θ		assume x
I	f^{oc}, A, B	f^{tr}, A, B		$\alpha_0 \longleftrightarrow q_0 \longleftrightarrow k$
$x \leqslant 1$; II	χ^{oc}, q^{oc}	χ^{tr}, q^{tr}		
			$\alpha_0, k, r_g, i, L_s, L_g$	

109. *β Lyrae type.* We take the case: circular orbit, oblate stars, uniform light. Let us consider two components which revolve around each other. If the stars are far apart, neither will be greatly distorted by the gravitational attraction of the other, but they may still have a rapid rotation around their axes, which causes an oblateness. These cases actually exist; the axes will be *a, a, c*. When the system is an eclipsing binary, the inclination will be close to 90° and the axes of rotation nearly in the plane of the sky. During the revolution we always see the same areas outside eclipse so that the maximum is flat.

Now let the components come nearer together in space (*β* Lyrae type). The tidal action becomes important, just as in the moon-earth system. This effect alone will deform the stars, giving them axes *a, b, b* with the major axes pointing toward each other so that they can be visualized as oblate stars.

The combination of tidal attraction and rotation produces axes a, b, c, all unequal. Theoretically it follows that there is the following relation between the axes:

$$b - c = \tfrac{2}{3}(a - b) \tag{312}$$

The real star may even deviate from this ellipsoid and be an egg-like figure. However, this is too difficult for computation, and we choose a more regular model, namely an ellipsoid of rotation or oblate star with axes a, b, b which comes as close as possible to the expected case. This at least gives a solution. The differences in the observed light curve may be called perturbations. If we compute a theoretical star according to formula (312), and compare it to the model we will actually use, we may find the following. Notice that the b of the model is intermediate between the b and c of the theoretical star.

theoretical	model
$a = 1.00$	$a = 1.00$
$b = 0.91$	$b = 0.88$
$c = 0.85$	$\epsilon = 0.12$

We can define the oblateness of the ellipsoid of rotation in much the same way as is done for the earth:

$$\epsilon = \frac{a - b}{a} = 1 - \frac{b}{a} \tag{313}$$

Often the ellipticity of the equatorial section is used:

$$e^2 = \frac{a^2 - b^2}{a^2} = 1 - \frac{b^2}{a^2} = \left(1 - \frac{b}{a}\right)\left(1 + \frac{b}{a}\right) \approx 2\epsilon \tag{314}$$

The ellipticity is found as the mathematical result of the computations, but it may easily be confused with the eccentricity of the orbit. Though the oblateness is an approximation we will use it, since it can be visualized more easily. Since in practice the oblateness is rarely more than 0.1 we can approximate in this way in all cases and neglect the ϵ^2.

110. *Oblateness*. For simplicity we take both components similar, and thus with the same ratio of axes. Sections through the centers give ellipses of the same ellipticity. Even without an eclipse these stars show disks, the areas of which change continuously during the period. Even if the inclination is such that no eclipse occurs, there is still a periodic light variation caused by the oblateness of the components. We will first study this type of light variation, because there are light curves caused in this way. Only later will we study the additional influence of the eclipse.

In the spherical triangle ABS_2 shown in Figure 135 in which $\angle A = 90°$ we have:

$$\cos \varphi = \sin i \cos \theta \tag{315}$$

Here θ is in the orbital plane, and φ is the angle between the line of sight and the line connecting the centers of the stars in space. We will always start with φ in the plane BC_1S_2, afterwards replace φ by θ and multiply by the factor $\sin i$ to get the result desired.

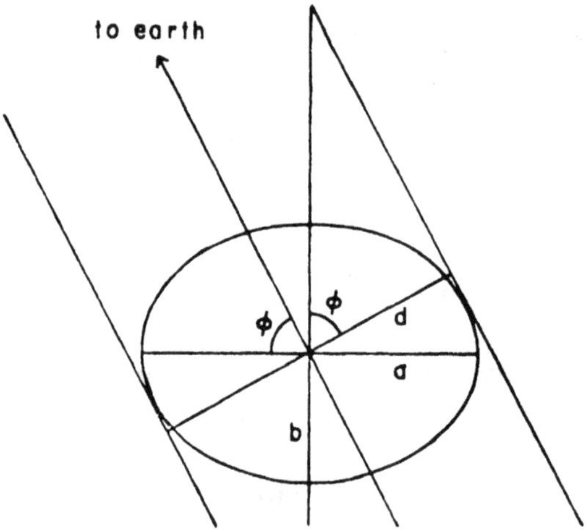

Figure 147. Relation between d and a as a function of φ.

Instead of the true ellipse a, b we observe the ellipse d, b. We first must find the connection between d and a as a function of φ. This is a mathematical problem. Consider the equations of the ellipse and a straight line (Figure 147):

$$\frac{x^2}{a^2} + \frac{y^2}{b^2} = 1$$

$$y = px + q, \qquad \text{where } p = \tan\varphi, \qquad q\cos\varphi = d$$

We find the intersections of the two from the roots of the equation:

$$(b^2 + a^2p^2)x^2 + (a^2 2pq)x + (a^2q^2 - a^2b^2) = 0$$

For this line to be tangent to the ellipse the discriminant must be zero:

$$(a^2 2pq)^2 - 4(b^2 + a^2p^2)(a^2q^2 - a^2b^2) = 0$$

$$-b^2q^2 + b^4 + a^2b^2p^2 = 0$$

$$b^2 + a^2p^2 = q^2, \qquad b^2 + a^2\tan^2\varphi = \frac{d^2}{\cos^2\varphi}$$

$$d^2 = a^2\sin^2\varphi + b^2\cos^2\varphi \tag{316}$$

Introducing the ellipticity we will rewrite this relation:

$$d^2 = a^2 - a^2\cos^2\varphi + b^2\cos^2\varphi = a^2 - (a^2 - b^2)\cos^2\varphi$$

$$= a^2 - a^2e^2\cos^2\varphi = a^2(1 - e^2\cos^2\varphi) = a^2(1 - e^2\sin^2 i\cos^2\theta)$$

Introducing the oblateness and using the well known relation:

$2\cos^2\theta = 1 + \cos 2\theta$, we have:

$$d^2 = a^2(1 - 2\epsilon\sin^2 i\cos^2\theta) = a^2(C - \epsilon\sin^2 i\cos 2\theta) \tag{317}$$

Often for abbreviation we use:

$$z = e^2\sin^2 i = 2\epsilon\sin^2 i \tag{318}$$

The maximum value of d is a; this occurs for φ or $\theta = 90°$ or 270°. The minimum value of d follows for φ or $\theta = 0°$ or 180°.

$$d_0^2 = a^2(1 - 2\epsilon\sin^2 i), \qquad d_0 = a(1 - \epsilon\sin^2 i)$$

Thus at maximum we see the normal ellipse a, b; during minimum we have the ellipse d_0, b which for $i = 90°$ becomes a circle b, b.

The uniform light of the components is proportional to the areas visible outside eclipse, and therefore we have:

$$\frac{l_s}{L_s} = \frac{l_g}{L_g} = \frac{\pi db}{\pi ab} = \frac{d}{a} = \sqrt{(1 - z\cos^2\theta)} = \sqrt{(1 - 2\epsilon\sin^2 i\cos^2\theta)}$$

$$= 1 - \epsilon\sin^2 i\cos^2\theta = C' - \tfrac{1}{2}\epsilon\sin^2 i\cos 2\theta$$

For the total light of both components we find:

$$l = l_s + l_g = (1 - \epsilon\sin^2 i\cos^2\theta)(L_s + L_g)$$

If again we take $L_s + L_g = 1$, or $l_{90} = 1$, we have:

$$l = 1 - \epsilon\sin^2 i\cos^2\theta = l_{45} - \tfrac{1}{2}\epsilon\sin^2 i\cos 2\theta \qquad (319)$$

$$l^2 = 1 - 2\epsilon\sin^2 i\cos^2\theta = C - \epsilon\sin^2 i\cos 2\theta$$

Without any approximation we find:

$$l^2 = 1 - z\cos^2\theta = C - \tfrac{1}{2}z\cos 2\theta \qquad (320)$$

The light curve around maximum has the form of a cosine curve with double argument, thus having two maxima and two minima during the period. Now we can plot l against $\cos^2\theta$ for the observations around the maxima (Figure 148). We get a linear relation, and the tangent of the slope gives us $\epsilon\sin^2 i$. If l is

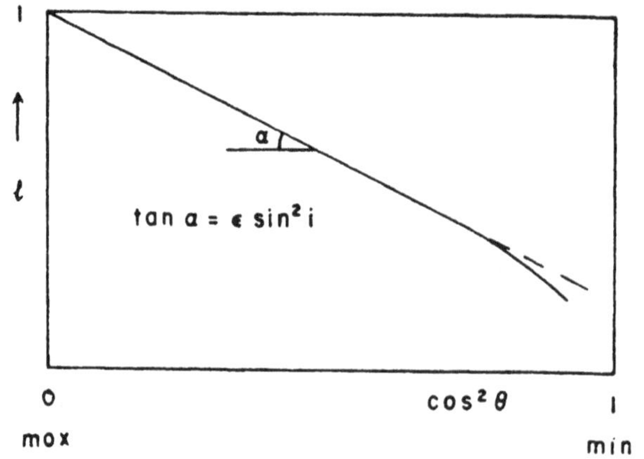

Figure 148. Plot of the intensity l in the maxima against $\cos^2\theta$.

plotted against cos 2θ, the tangent of the slope gives $\frac{1}{2}\,\epsilon\sin^2 i$. Plotting l^2 against $\cos^2\theta$ gives $z = e^2\sin^2 i = 2\epsilon\sin^2 i$, and perhaps this is the best way of proceeding. To start, we can take $i = 90°$ and find a provisional ϵ. In this way the oblateness can be found.

111. *Dynamical condition for oblate stars.* During a minimum the observed light is affected by both oblateness and the eclipse. All minima are therefore curved.

$$l = l_g + (1 - \alpha)\,l_s = 1 - \alpha l_s$$
$$l = (1 - \epsilon\sin^2 i\cos^2\theta)\,\{L_g + (1 - \alpha)\,L_s\}$$
$$l = (1 - \epsilon\sin^2 i\cos^2\theta)\,(1 - \alpha L_s) \qquad (321)$$

The first factor is caused by the oblateness effect; the second by the eclipse of the components. Expressed in magnitudes we get two terms:

$$m = -2.5\log(1 - \epsilon\sin^2 i\cos^2\theta) - 2.5\log(1 - \alpha L_s) \qquad (322)$$

We know $\epsilon\sin^2 i$ from the maxima; for each phase θ is known. We can compute $1 - \epsilon\sin^2 i\cos^2\theta = l_{45} - \frac{1}{2}\,\epsilon\sin^2 i\cos 2\theta \leqslant 1$ for each normal point. We will rectify the light curve, that is, remove the influence of the oblateness. This is a process of division in the case of intensities and a process of subtraction in the case of magnitudes (Figure 149).

$$l_r = \frac{l}{1 - \epsilon\sin^2 i\cos^2\theta} = \frac{l}{1 - \frac{1}{2}z\cos^2\theta} \qquad (323)$$

The rectified light curve has flat maxima, $\bar{l}_{r,\,\max} = 1$. For any phase the ellipse has axes a, b in this artificial light curve. We are allowed to multiply the intensity by a factor a/b, in which case we consider a circular disk with axes a, a.

The minima can be analyzed as before, but let us see how the eclipse occurs in reality. We actually have eclipses of disks of changeable size, and not eclipses of disks with axes a, b or of circular disks a, a. This means that even after rectification of the

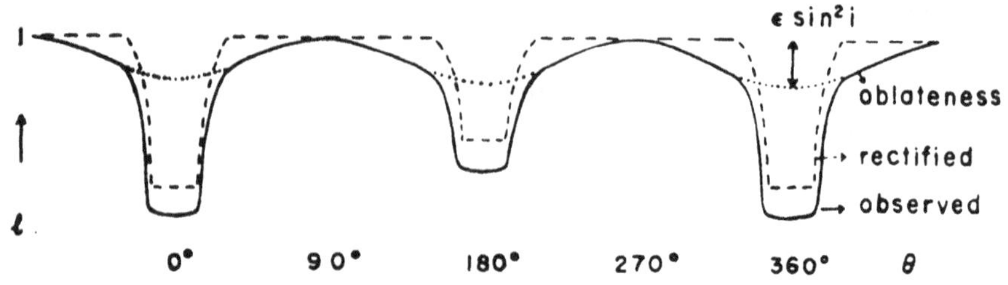

Figure 149. The maxima of the observed light curve are curved by oblateness. The rectified light curve shows constant maxima. The rectified minima can now be studied as before.

intensity there remains some complication. The covered area of the ellipse *d*, *b* is expressed in terms of the small ellipse as a unit according to the definitions. Let us now consider the case for $i = 90°$. By multiplying all lengths in the y direction by the same factor d/b we get auxiliary circles. The covered area increases in the same ratio as the area of the ellipse, so that the new covered area is effectively the same as before, when expressed in terms of the auxiliary circle *d*, *d*. This holds also for $i \neq 90°$. We can now work with the auxiliary circles of changing radii d_g and $d_s = k\, d_g$ for which the old dynamical condition holds (Figure 150).

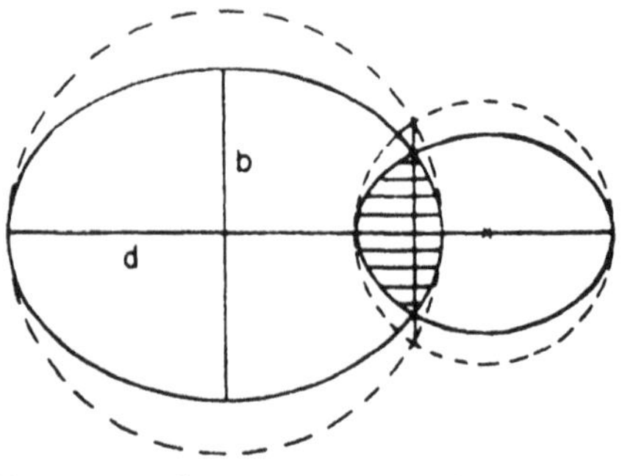

Figure 150. The covered area of the ellipses increases in the same ratio as the area of the auxiliary circles.

Method I gives:

$$\left(\frac{\delta}{d_g}\right)^2 = \frac{\cos^2 i}{d_g^2} + \frac{\sin^2 i}{d_g^2}\sin^2\theta = f^2(k, \alpha) \tag{324}$$

where $\delta/d_g = f(k, \alpha)$ is given in the previously mentioned table.

$$d^2 = a^2(1 - 2\epsilon\sin^2 i\cos^2\theta) = a^2(1 - z\cos^2\theta) = a^2E(\theta)$$

Substitution gives:

$$\left(\frac{\delta}{a_g}\right)^2\frac{1}{E(\theta)} = \frac{\cos^2 i}{a_g^2E(\theta)} + \frac{\sin^2 i}{a_g^2E(\theta)}\sin^2\theta = f^2(k, \alpha)$$

$$\left(\frac{\delta}{a_g}\right)^2 = \frac{\cos^2 i}{a_g^2} + \frac{\sin^2 i}{a_g^2}\sin^2\theta = f^2(k, \alpha)(1 - z\cos^2\theta) \tag{325}$$

This correction factor depending on phase angle is the complication mentioned above. It is caused by the eclipses really taking place on disks of changing size. After applying this correction factor to $f^2(k, \alpha)$, we find k, a_g, i from the branches in the same way as before. Thus we find the semi-major axis. With i found in this way, and $z = 2\epsilon\sin^2 i$ known from the maxima, we find ϵ.

Method II requires that the phase be rectified in order that we be able to use the old formulae. The proof is rather laborious, and we will give only an outline. The dynamical condition is:

$$\delta^2 = \cos^2 i + \sin^2 i\sin^2\theta = d_g^2(1 + kp)^2 = a_g^2(1 - z\cos^2\theta)(1 + kp)^2$$

$$a_g^2(1 + kp)^2 = \frac{\cos^2 i + \sin^2 i\sin^2\theta}{1 - z\cos^2\theta} \tag{326}$$

For two points on the curve we get:

$$a_g^2\{(1 + kp_1)^2 - (1 + kp_2)^2\} = \frac{\cos^2 i + \sin^2 i\sin^2\theta_1}{1 - z\cos^2\theta_1} - \frac{\cos^2 i + \sin^2 i\sin^2\theta_2}{1 - z\cos^2\theta_2}$$

Working this out and introducing sines everywhere we have:

$$a_g^2\{(1 + kp_1)^2 - (1 + kp_2)^2\} = \frac{(\sin^2\theta_1 - \sin^2\theta_2)(\sin^2 i - z)}{(1 - z\cos^2\theta_1)(1 - z\cos^2\theta_2)}$$

We can write a similar expression for the second and third points and then divide. The procedure is thus quite the same as we have had before. We find:

$$\frac{(1+kp_1)^2 - (1+kp_2)^2}{(1+kp_2)^2 - (1+kp_3)^2} = \frac{(\sin^2\theta_1 - \sin^2\theta_2)(1 - z\cos^2\theta_3)}{(\sin^2\theta_2 - \sin^2\theta_3)(1 - z\cos^2\theta_1)} = \psi(k, \alpha)$$

Now we compute:

$$\frac{\sin^2\theta_1}{1 - z\cos^2\theta_1} - \frac{\sin^2\theta_2}{1 - z\cos^2\theta_2} = \frac{(\sin^2\theta_1 - \sin^2\theta_2)(1 - z)}{(1 - z\cos^2\theta_1)(1 - z\cos^2\theta_2)}$$

Here again we have introduced sines everywhere in the numerator. Thus we have now:

$$a_g^2\{(1+kp_1)^2 - (1+kp_2)^2\} = \\ = \left\{\frac{\sin^2\theta_1}{1 - z\cos^2\theta_1} - \frac{\sin^2\theta_2}{1 - z\cos^2\theta_2}\right\}\left(\frac{\sin^2 i - z}{1 - z}\right)$$

$$\frac{(1+kp_1)^2 - (1+kp_2)^2}{(1+kp_2)^2 - (1+kp_3)^2} = \psi(k, \alpha)$$

$$= \left\{\frac{\sin^2\theta_1}{1 - z\cos^2\theta_1} - \frac{\sin^2\theta_2}{1 - z\cos^2\theta_2}\right\} : \left\{\frac{\sin^2\theta_2}{1 - z\cos^2\theta_2} - \frac{\sin^2\theta_3}{1 - z\cos^2\theta_3}\right\}$$

From this formula we see that the rectification in phase is:

$$\sin^2\theta_r = \frac{\sin^2\theta}{E(\theta)} = \frac{\sin^2\theta}{1 - z\cos^2\theta} \tag{327}$$

Let us define the constants:

$$A_r = \frac{\sin^2\theta_2}{1 - z\cos^2\theta_2}, \qquad B_r = A_r - \frac{\sin^2\theta_3}{1 - z\cos^2\theta_3} \tag{328}$$

The formula reduces to the old form:

$$\frac{\sin^2\theta_r - A_r}{B_r} = \psi(k, \alpha) \qquad \text{For } \theta = 0,\ \psi(k, \alpha_0) = -\frac{A_r}{B_r}$$

$$\sin^2\theta_r = A_r + B_r\,\psi(k, \alpha) \tag{329}$$

This means that after rectification in both intensity and phase we can proceed in the old way. For partial eclipses we find similarly:

$$\sin^2\theta_r(n) = \sin^2\theta\left(\tfrac{1}{2}\right).\,\chi(k, n, \alpha_0) \tag{330}$$

Also the dynamical condition can be written in similar form:

$$1 - z\sin^2\theta_r = 1 - \frac{z\sin^2\theta}{1 - z\cos^2\theta} = \frac{1-z}{1 - z\cos^2\theta}$$

$$\frac{1 - z\sin^2\theta_r}{1-z} = \frac{1}{1 - z\cos^2\theta}$$

Substituting into the dynamical condition we find:

$$\begin{aligned}
a_g^2(1+kp)^2 &= \frac{\cos^2 i}{1 - z\cos^2\theta} + \frac{\sin^2 i\sin^2\theta}{1 - z\cos^2\theta} \\
&= \cos^2 i\,\frac{(1 - z\sin^2\theta_r)}{1-z} + \sin^2 i\sin^2\theta_r \\
&= \frac{\cos^2 i}{1-z} - \frac{z\cos^2 i\sin^2\theta_r}{1-z} + \sin^2 i\sin^2\theta_r \\
&= \frac{\cos^2 i}{1-z} + \frac{(1-z)\sin^2 i - z\cos^2 i}{1-z}\sin^2\theta_r \\
&= \frac{\cos^2 i}{1-z} + \frac{\sin^2 i - z}{1-z}\sin^2\theta_r
\end{aligned}$$

Let us call:

$$\left.\begin{aligned}
\cos^2 i_r &= \frac{\cos^2 i}{1-z} \\
\sin^2 i_r &= 1 - \frac{\cos^2 i}{1-z} = \frac{1 - z - \cos^2 i}{1-z} = \frac{\sin^2 i - z}{1-z}
\end{aligned}\right\} \tag{331}$$

The dynamical condition appears in the same form:

$$\cos^2 i_r + \sin^2 i_r\sin^2\theta_r = a_g^2(1+kp)^2 \tag{332}$$

To summarize we have found the following rectification formulae:

$$\left.\begin{aligned}
&l_r = \frac{l}{1 - \frac{1}{2}z\cos^2\theta}, \qquad \text{flat maxima } \bar{l}_{r,\,\max} = 1 \\
&\text{(I)} \qquad \left(\frac{\delta}{a_g}\right)^2 = f^2(k, \alpha)(1 - z\cos^2\theta) \\
&\text{(II)} \qquad \sin^2\theta_r = \frac{\sin^2\theta}{1 - z\cos^2\theta}, \quad \cos^2 i_r = \frac{\cos^2 i}{1-z}
\end{aligned}\right\}$$

In reality the stars usually do not have similar forms. Then they can be transformed to an auxiliary circle and an auxiliary ellipse. The computation is much more complicated but can be carried through. This has been done by L. Plaut.

112. *Oblateness with limb darkening.* We consider the case: circular orbit, oblate stars, limb darkening. Again we are considering the light variation in the maxima, first of all for total limb darkening. Let $d\sigma$ be a surface element of the ellipsoid of rotation with direction cosines l, m, n with respect to the axes of the star. Further, the direction cosines of the line of sight on the same axes are λ, μ, ν. The γ is again the angle between the normal and the line of sight (Figure 151).

$$\cos\gamma = l\,\lambda + m\,\mu + n\,\nu$$

The apparent projected surface of the surface element $d\sigma$ is now $\cos\gamma\; d\sigma$. We have seen earlier that for total limb darkening we perceive $I_0 \cos\gamma$ per unit area. From the surface element we thus observe the following intensity:

$$I_0 \cos\gamma \,.\, \cos\gamma\; d\sigma = I_0 \cos^2\gamma\; d\sigma$$

The total intensity of the star as seen by the observer is:

$$\text{int.} = \int_{\frac{1}{2}\,\text{ell.}} I_0 \cos^2\gamma\; d\sigma = \tfrac{1}{2} I_0 \int_{\text{ell.}} \cos^2\gamma\; d\,\sigma$$

$$= \tfrac{1}{2} I_0 \int_{\text{ell.}} (l\,\lambda + m\,\mu + n\,\nu)^2\, d\sigma$$

$$= \tfrac{1}{2} I_0 \int_{\text{ell.}} (l^2\lambda^2 + m^2\mu^2 + n^2\nu^2)\, d\sigma + \int_{\text{ell.}} (\text{cross products})$$

We want to prove that the integrals of the cross products are all zero by reason of symmetry. For a given time λ, μ, ν are constant, but l, m, n change depending on the surface element chosen. Keep l constant and choose some m; diametrically

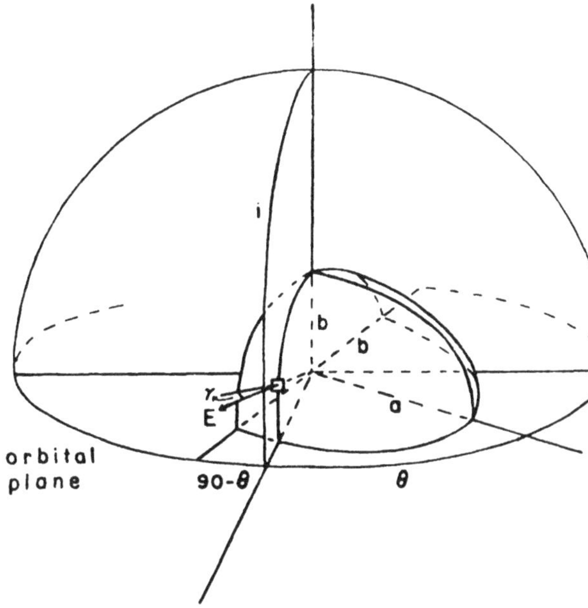

Figure 151. An oblate star with total limb darkening.

opposite there is always $-m$. The sum of these two is zero. Thus the cross product $(\lambda\mu)$ $(l\ m)$ forms a sum which equals zero for any instant. Because λ, μ, ν are constants for a given time we now have:

$$\text{int.} = \lambda^2 \tfrac{1}{2} I_0 \int l^2 d\sigma + \mu^2 \tfrac{1}{2} I_0 \int m^2 d\sigma + \nu^2 \tfrac{1}{2} I_0 \int n^2\, d\sigma \qquad (333)$$

Now $l^2 d\sigma$ is constant for each moment by reason of symmetry. The surface elements with constant l are situated on a ring of the ellipsoid; the influence of all rings is constant. From the spherical triangles with 90° angles we see:

$$\left.\begin{aligned} \lambda &= \sin i \cos \theta \\ \mu &= \sin i \sin \theta \\ \nu &= \cos i \end{aligned}\right\}$$

These depend only on θ, thus on the time. Further define the constants:

$$\left.\begin{aligned} P &= \tfrac{1}{2} I_0 \int l^2 d\sigma \\ Q &= \tfrac{1}{2} I_0 \int m^2 d\sigma \\ R &= \tfrac{1}{2} I_0 \int n^2 d\sigma \end{aligned}\right\} \qquad (334)$$

P is related to l thus also to a; Q is related to m, thus to b. Thus $(Q - P)$ is consequently related to $(a - b)$, therefore to the oblateness ϵ. If we again call the intensity l, we find:

$$l = P \sin^2 i \cos^2\theta + Q \sin^2 i \sin^2\theta + R \cos^2 i \qquad (335)$$

$$\text{For } \theta = 90°;\ l_{max} = Q \sin^2 i \qquad + R \cos^2 i$$

After subtraction we have the following result:

$$\begin{aligned} l - l_{max} &= P \sin^2 i \cos^2\theta + Q \sin^2 i (\sin^2\theta - 1) \\ &= P \sin^2 i \cos^2\theta - Q \sin^2 i \cos^2\theta = (P - Q) \sin^2 i \cos^2\theta \end{aligned}$$

$$l = l_{max} + (P - Q) \sin^2 i \cos^2\theta = 1 - (Q - P) \sin^2 i \cos^2\theta$$

This is of the form:

$$l = 1 - f(\epsilon) \sin^2 i \cos^2\theta = l_{45} - \tfrac{1}{2} f(\epsilon) \sin^2 i \cos 2\theta \qquad (336)$$

We have still to determine $f(\epsilon) = Q - P$. These values Q and P can be computed, but in a rather complicated way; A. J. Wesselink has derived them more simply. For $i = 90°$ we have:

$$l_{min} = 1 - f(\epsilon), \quad l_{max} = 1, \qquad \frac{l_{min}}{l_{max}} = 1 - f(\epsilon) \qquad (337)$$

Compute the intensity from the minimum areas of the stars and express this in terms of the intensity radiated by the maximum areas. The ratio gives $1 - f(\epsilon)$. To find the whole light variation around the maximum for the general case of $i \neq 90°$ we have only to multiply $f(\epsilon)$ by $\sin^2 i \cos^2\theta$.

(1) For $i = 90°$ and $\theta = 0°$ we see circles as individual areas. The surface brightness is the same in each of the concentric circular rings, but the circular disk is here the projection of an ellipsoid and not of a sphere. The situation therefore differs

somewhat from the conventional case. Taking the sum of the intensities of all rings, or better integrating over all the rings, gives the total intensity received. Dividing by the area of the circle gives the mean surface brightness (Figure 152):

$$\bar{I}_{\min} = \tfrac{2}{3}\,(1 - \tfrac{2}{5}\,\epsilon) \qquad (338)$$

The factor $\frac{2}{3}$ was caused by the total limb darkening, so $-\frac{2}{5}\,\epsilon$ shows the influence of the oblateness.

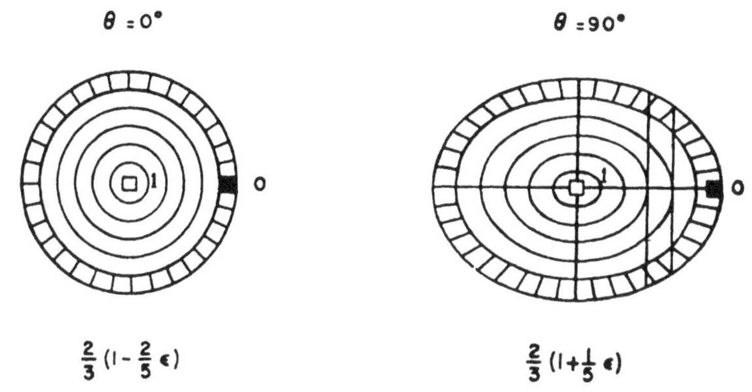

Figure 152. Mean surface brightness of an oblate component with total limb darkening at minimum and maximum light respectively ($i = 90°$).

(2) For $i = 90°$ and $\theta = 90°$ we see the stars as ellipses. In each of the concentric elliptical rings around the center there is constant surface luminosity. Adding over all rings would give the desired integral. But there is also another method, namely to break up the surface area into strips each of which suffers from the darkening effect. We can therefore perform the double integration: first summing up one strip, then summing over all strips of the whole ellipse. For the mean surface brightness we find:

$$\bar{I}_{\max} = \tfrac{2}{3}\,(1 + \tfrac{1}{5}\epsilon) \qquad (339)$$

The influence of the oblateness is thus $+\frac{1}{5}\epsilon$. The area of the ellipse is larger than that of the circle by a factor:

$$\frac{\text{area ellipse}}{\text{area circle}} = \frac{\pi ab}{\pi bb} = \frac{a}{b} = \frac{1}{1-\epsilon} = 1 + \epsilon$$

For the ratio of minimum light to maximum light we have:

$$\frac{l_{\min}}{l_{\max}} = \frac{1}{(1+\epsilon)}\frac{(1-\frac{2}{5}\epsilon)}{(1+\frac{1}{5}\epsilon)} = \frac{1-\frac{2}{5}\epsilon}{1+\frac{6}{5}\epsilon}$$
$$=(1-\tfrac{2}{5}\epsilon)(1-\tfrac{6}{5}\epsilon) = 1 - \tfrac{8}{5}\epsilon$$

For total limb darkening we find the light variation in the maximum by the following expression (Figure 153):

$$l = 1 - \tfrac{8}{5}\epsilon \sin^2 i \cos^2\theta = l_{45} - \tfrac{4}{5}\epsilon \sin^2 i \cos 2\theta \qquad (340)$$

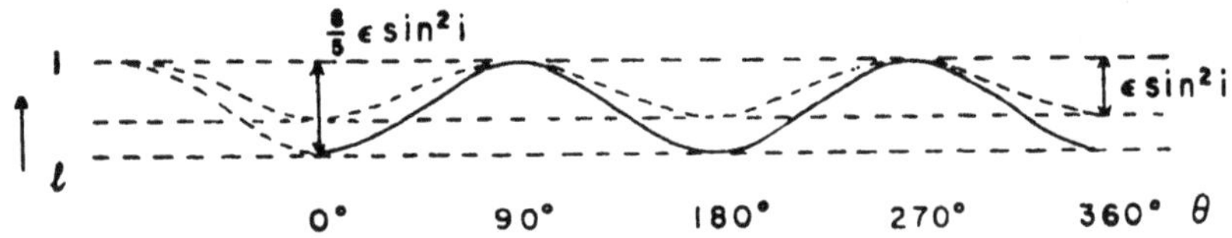

Figure 153. The amplitude of the light variation in the maxima as caused by oblateness is 1.6 times as large for total limb darkening as for uniform light.

The limb darkening caused the amplitude of the light variation to increase by a factor 1.6 compared with that of uniform light.

For intermediate darkening we make use of the light balance.

$$\frac{l_{\min}}{l_{\max}} = \frac{1}{(1+\epsilon)} \cdot \frac{(1-x) + x\frac{2}{3}(1-\frac{2}{5}\epsilon)}{(1-x) + x\frac{2}{3}(1+\frac{1}{5}\epsilon)}$$
$$= \frac{1}{(1+\epsilon)}\frac{(1-\frac{1}{3}x-\frac{4}{15}x\epsilon)}{(1-\frac{1}{3}x+\frac{2}{15}x\epsilon)} = \frac{1-\frac{1}{3}x-\frac{4}{15}x\epsilon}{1-\frac{1}{3}x+\epsilon-\frac{1}{5}x\epsilon}$$
$$= \frac{15-5x-4x\epsilon}{15-5x+15\epsilon-3x\epsilon} = 1 - \frac{15+x}{15-5x}\epsilon$$

For intermediate limb darkening we therefore find:

$$l = 1 - \frac{15 + x}{15 - 5x}\,\epsilon \sin^2 i \cos^2\theta = l_{45} - \frac{15 + x}{30 - 10x}\,\epsilon \sin^2 i \cos 2\theta \qquad (341)$$

The left factor is thus 1.0 for uniform light and 1.6 for total limb darkening and between these values for intermediate limb darkening. For example, for $x = \frac{1}{2}$ we find the factor 1.24 (Figure 154). From the observations around the maxima we again plot l against $\cos^2\theta$ or some similar graph. This gives us the value of $f(\epsilon)\sin^2 i$ which we have to use to rectify the curve. From the branch of the rectified light curve we find a_g, k, i, x. From this x we can find the value of the factor; with i known, ϵ follows.

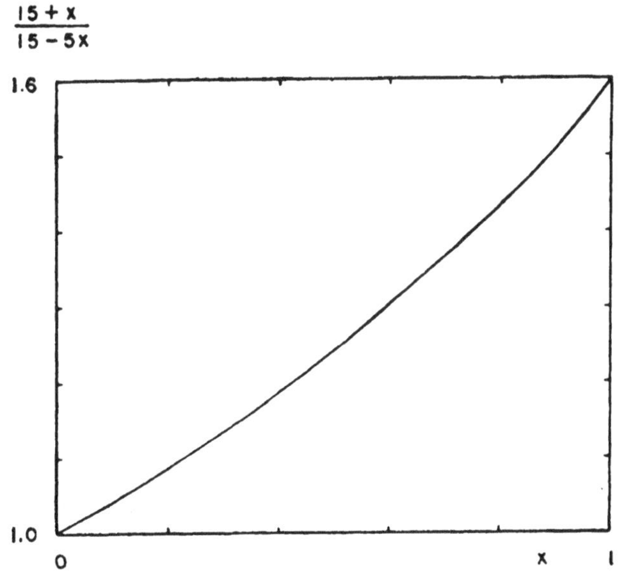

Figure 154. The relation between the limb darkening x and the factor in the amplitude of the light variation in the maxima.

113. *W Ursae Majoris type.* Here the components are so close that they almost touch each other. The periods are of the order of half a day, the eclipses last for the longest part of the periods. In the plot of l against $\cos^2\theta$ or a similar one, only a short part near $\theta = 90°$ is unaffected by the eclipse, and is caused by oblateness alone. In this plot the tangent to the curve at $\cos^2\theta = 0$ gives the desired $f(\epsilon)\sin^2 i$. The rest of the orbital determination is as usual.

There is a simple relation between period and mean stellar density. Let us take spherical stars of radii r_1 and r_2. The density is ρ and the mass $\mathfrak{M}$. In addition we need Kepler's third law, in which R is the radius of the orbit.

$$\mathfrak{M} = \rho\tfrac{4}{3}\pi r^3, \qquad \mathfrak{M}_1 + \mathfrak{M}_2 = \frac{R^3}{P^2}$$

The units are here solar mass, astronomical unit, and year.

$$\frac{\rho_1 r_1^3 + \rho_2 r_2^3}{\rho_\odot r_\odot^3} = \frac{R^3}{P^2}$$

Now take the solar radius as the unit for r_1, r_2, R, the solar density as the unit for density, and P in days. We have then because 1 astronomical unit = 215 solar radii:

$$\rho_1 r_1^3 + \rho_2 r_2^3 = \left(\frac{R}{215}\right)^3 \left(\frac{365.25}{P}\right)^2 = \frac{R^3}{74.5\,P^2}$$

In case the components are exactly the same and touching each other we have $\rho = \rho_1 = \rho_2$ and $r = r_1 = r_2 = R/2$.

$$2\rho\left(\frac{r}{R}\right)^3 = \frac{1}{74.5\,P^2}, \qquad 18.6\,P^2\rho = 1$$

For an ellipsoid we use ab^2 instead of r^3, thereby decreasing the coefficient to approximately 16. Thus for touching oblate stars:

$$16\,P^2\rho = 1, \qquad P\sqrt{\rho} = \tfrac{1}{4} = 0.25 \qquad (342)$$

In logarithmic form we get:

$$2\log P + \log\rho = -1.2$$

This value defines the lower limit. In Figure 155 the W Ursae

Majoris stars, the β Lyrae stars, and the Algol stars are respectively situated more and more to the right of this line. It is interesting that the theory of stellar interiors gives a similar relation for pulsating stars with a nearly constant product. The observations show that for cluster variables $P\sqrt{\rho} = 0.1$ and for Cepheids $P\sqrt{\rho} = 0.15$. This relation is quite independent of that for eclipsing binaries.

The above relation gives a quick estimate of the mean density of the stars expressed in terms of the sun's density, which is 1.41 times that of water; but this is only a lower limit. We proceed as follows:

$$1 + \frac{\rho_2 r_2^3}{\rho_1 r_1^3} = \frac{R^3}{74.5\,P^2} \cdot \frac{1}{\rho_1 r_1^3}$$

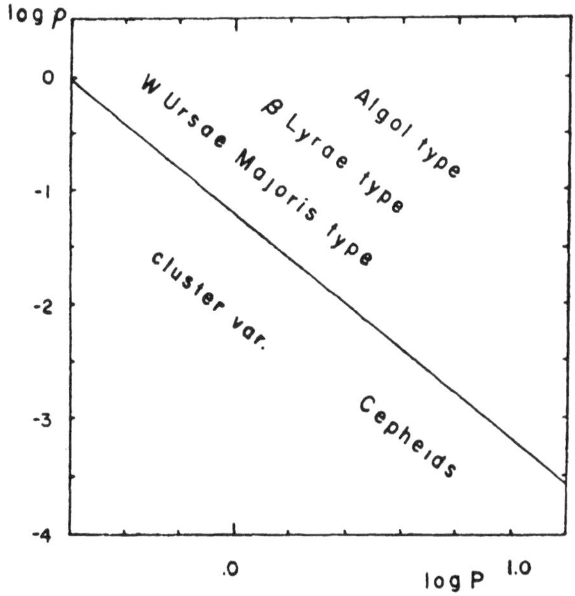

Figure 155. The relation between the logarithm of the period and the logarithm of the mean density for different types of variables.

We can introduce the mass ratio of the components.

$$1 + \frac{\mathfrak{M}_2}{\mathfrak{M}_1} = \frac{1}{74.5\, P^2 \rho_1} \left(\frac{R}{r_1} \right)^3$$

We notice here that we can use our old definition of r_1 in terms of $R = 1$. Usually this expression is written in the form:

$$P^2 \rho_1 = \frac{1}{74.5\, r_1^3 \left(1 + \frac{\mathfrak{M}_2}{\mathfrak{M}_1} \right)} \tag{343}$$

For oblate stars we replace r^3 by ab^2. The relation holds just as well for elliptical orbits if we take the semi-major axis as a unit.

$$P^2 \rho_1 = \frac{1}{74.5\, a_1 b_1^2 \left(1 + \frac{\mathfrak{M}_2}{\mathfrak{M}_1} \right)} \tag{344}$$

A similar expression for the other star is found by interchanging the subscripts 1 and 2. The mass ratio follows from spectroscopic observations. We have found r_1 and r_2 in terms of R.

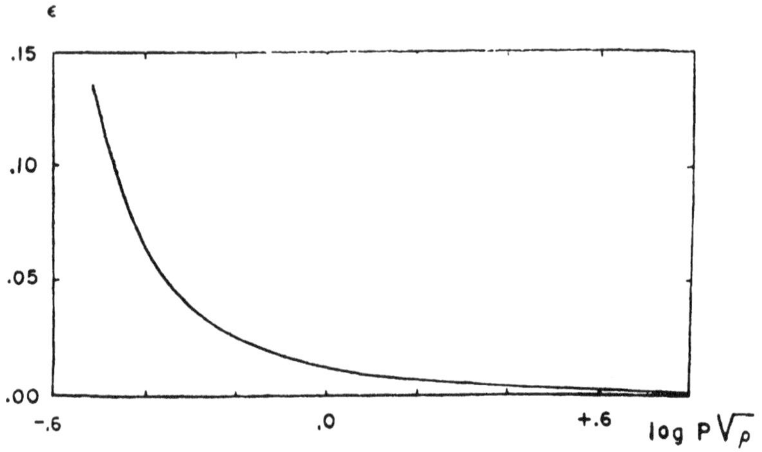

Figure 156. The relation between the oblateness and the logarithm of $P\sqrt{\rho}$ (P = period, ρ = mean density).

When the oblateness is known, we can also find the ab^2 in terms of R^3. In addition P is known so that ρ_1 follows. For the other component P is the same, but ρ_2 is usually different. Therefore both points are in a vertical line. See Figure 156 for the relation between the oblateness and $\log P \sqrt{\rho}$.

W Ursae Majoris stars are so close to each other that we can expect much interaction. The periods are not strictly constant; during the maxima there are also strange changes in the light which may be caused by huge prominences and flares. Both minima are observed to be redder than the maxima. In both cases we are looking at the cooler and redder ends of the oblate stars during such a minimum as a consequence of the gravitation and reflection effects.

114. *Kinds of eclipses.* A typical W Ursae Majoris system has very similar components. Two spectra are therefore visible. However, the radial velocity curves are as a rule different from a more normal system and this makes an interesting conclusion possible. The primary minimum or phase $\theta = 0$ falls on the middle of the increasing branch of the radial velocity curve with the smallest amplitude and corresponding to the strongest line in the spectrum. (See Figure 92.) The star in front during primary minimum is therefore the most massive one because just afterwards this star is receding. The ratio of intensities of the components can be found as ratio of line depths or as the ratio of the equivalent widths for these similar spectral types (see paragraph 67). The star in front is therefore the brightest one because it gives the stronger lines in the spectrum. Moreover, we know from the light curve that at primary minimum the star in front has the smallest surface luminosity and is therefore the cooler. This is correct for circular orbit and uniform light or similar limb darkening.

Let the component in front during primary minimum have

the subscript one and let $\bar{I}$ be the mean surface luminosity. For $i = 90°$ and $\theta = 0$ the a, b, b ellipsoid gives a circular disk with radius b.

$$l_1 = \pi b_1^2 \bar{I}_1 > l_2 = \pi b_2^2 \bar{I}_2 \tag{345}$$

But $\bar{I}_1 < \bar{I}_2$ which is only possible if $b_1 > b_2$ because the negative sign has no sense here. Therefore the star in front is also the largest and we have an occultation at $\theta = 0$.

For $i \neq 90°$ we see during the minima an ellipse with axes d_0, b, but even for an extreme case like $i = 70°$ and $\epsilon = 0.1$ the disks are practically circular, and it is found that $d_0 = 1.013\ b$. We can replace b^2 by $d_0 b$ in the above inequality and the conclusion remains the same for our working model. A study of the shape of the minima gives independent information about the type of eclipse and these results agree for the accurately observed light curves proving that the real star does not deviate greatly from our model. Because the components are similar in size the k is usually found near unity and the information found is therefore valuable.

It can be seen that for the radial velocity curves of a more normal system such a conclusion cannot be reached. However, very few W Ursae Majoris stars show such a normal behavior.

115. *Gravitation effect.* We consider the case: circular orbit, oblate stars, gravitation effect (thus except for this last effect, uniform light). Let $\mathfrak{M}$ be the mass of the star and d the distance from the center to a mass point $\mathfrak{M}'$ on the surface. According to the law of gravitation:

$$F = G\frac{\mathfrak{M}\,\mathfrak{M}'}{d^2} = \mathfrak{M}' g, \qquad g = G\frac{\mathfrak{M}}{d^2}$$

Here F is the force, G the gravitational constant and g the acceleration. If we assume the energy to come from a point source in the center of the star, the luminosity of a star when it leaves the surface is also proportional to $1/d^2$. Therefore:

$$I \propto \frac{1}{d^2} \propto g \propto T^4$$

If we take I_0 as the unit of surface luminosity and g_0 as the corresponding unit of gravitational acceleration, we have:

$$\frac{I}{I_0} = \frac{g}{g_0} = 1 - \left(1 - \frac{g}{g_0}\right)$$

This holds for bolometric light. For photographic light we have:

$$\frac{I}{I_0} = 1 - y\left(1 - \frac{g}{g_0}\right)$$

$$I = I_0\left(1 - y + y\frac{g}{g_0}\right) \tag{346}$$

This is a formula similar to the one we have found for limb darkening. For $g = g_0$ we see that $I = I_0$. For a spherical star g is the same at every point of the surface and the surface luminosity caused by this effect is therefore uniform. It can be seen that the gravitation effect we are considering is a relative effect.

However, we shall express the gravitation effect somewhat differently. The minimum surface brightness appears at the two ends of the oblate star, and we shall take this as unity. The maximum occurs just between the ends on a circle of radius b, and this surface luminosity we shall call $1 + q\epsilon$. Thus $q\epsilon$ is the so-called gravitation effect, which for small oblateness is proportional to the oblateness. For a point in between on the surface the brightness is $1 + q\epsilon \sin^2\varphi$; therefore, between 1 and $1 + q\epsilon$. The $\sin \varphi$ would not work because the result has to be the same for negative angles; $\cos \varphi$ or $\cos^2\varphi$ are ruled out from the start.

We can contrast the gravitation effect on an oblate star with that of limb darkening on the same star. During minimum for $i = 90°$ the point of the oblate star is in the center of the circular disk. Then the gravitation effect makes the star brighter towards the edge, but limb darkening works just the opposite, so that we have to find opposite signs for the two effects. During maximum both effects help each other, and we must find identical signs.

We shall compute now the influence of the gravitation effect upon the maximum of the light curve in the same way as before.

(1) In the minimum we see a circle for $i = 90°$. In each ring we have uniform surface luminosity, $1 + q\,\epsilon \sin^2\varphi$. We have to take into account that, in general, on an ellipsoid the radius vector and the normal to the surface are not collinear. Multiplying by the area of the ring gives the intensity of the ring; adding the intensities of all rings, or better integrating over the whole disk, gives the intensity of the star. Dividing by the area of the star determines the area surface brightness, which comes out exactly half way between the extremes (Figure 157).

$$\bar{I}_{\text{min}} = 1 + \tfrac{1}{2} q\,\epsilon \tag{347}$$

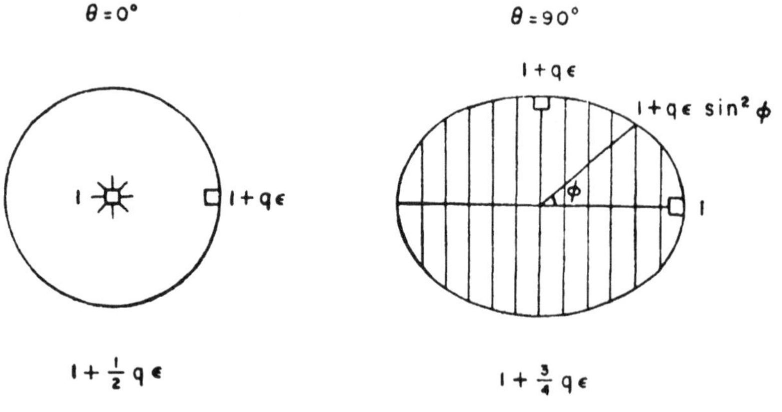

Figure 157. Mean surface brightness of an oblate component with gravitation effect at minimum and maximum light respectively ($i = 90°$).

(2) In the maximum we see an ellipse for $i = 90°$; we make strips as shown in the figure. For the mean surface brightness one finds now after the computation:

$$\bar{I}_{max} = 1 + \tfrac{3}{4} q \epsilon \tag{348}$$

We have found that the area of an ellipse is $(1 + \epsilon)$ times that of a circle. For bolometric light we have:

$$\frac{l_{min}}{l_{max}} = \frac{1 + \frac{1}{2} q\epsilon}{(1 + \epsilon)(1 + \frac{3}{4} q\epsilon)} = \frac{1 + \frac{1}{2} q\epsilon}{1 + \frac{3}{4} q\epsilon + \epsilon}$$

$$= 1 - \frac{\frac{1}{4} q\epsilon + \epsilon}{1 + \frac{3}{4} q\epsilon + \epsilon} = 1 - \left(1 + \frac{q}{4}\right)\epsilon$$

For photographic light this becomes:

$$\frac{l_{min}}{l_{max}} = 1 - \left(1 + \frac{hq}{4}\right)\epsilon = 1 - (1 + y)\,\epsilon \tag{349}$$

Here $h = f(\lambda, T)$ and can be computed for both a black and a grey body. For bolometric light $h = 1$. For photographic light the λ is constant, so that h depends only on the temperature of the star and falls in the range $0.25 \leqslant h \leqslant 1.25$. The q can be computed for different models; we shall consider the extreme cases. For a star of homogeneous density as in the MacLaurin model $q = 1$; for the Roche model in which all mass is concentrated at the center $q = 4$. This last model is closer to reality. For the entire range of wavelength we find therefore:

$$1.06 \leqslant 1 + \frac{hq}{4} = 1 + y \leqslant 2.25$$

For monochromatic light h is fixed, and the practical range of $(1 + y)$ is therefore smaller. Also, for a given spectral type we have some information about the model thus about q and the limb darkening x. As a result of such computations we can give q as a function of x.

The light variation in the maximum is:

$$l = 1 - (1 + y)\,\epsilon \sin^2 i \cos^2\theta = l_{45} - \tfrac{1}{2}(1 + y)\,\epsilon \sin^2 i \cos 2\theta \tag{350}$$

From the slope we again find $(1 + y)\,\epsilon \sin^2 i$. The i we find from the branch, and then determine $(1 + y)\,\epsilon$ from the combination of the observations. Theoretically y can be found as seen above. For two components similar to the sun, for which the density gradient of the interior is known, this is rather simple. Thus we find ϵ.

116. *Gravitation effect and limb darkening*. This case includes: circular orbit, oblate stars, limb darkening, gravitation. The surface brightness can now be written as:

$$I = I_0(1 - x + x\cos\gamma)\left(1 - y + y\frac{g}{g_0}\right) \tag{351}$$

We can write the result for total limb darkening at once:

$$\frac{l_{min}}{l_{max}} = 1 - \tfrac{8}{5}(1 + y)\,\epsilon$$

The intensity in the maximum is therefore:

$$l = 1 - \tfrac{8}{5}(1 + y)\,\epsilon \sin^2 i \cos^2\theta = l_{45} - \tfrac{4}{5}(1 + y)\,\epsilon \sin^2 i \cos 2\theta \tag{352}$$

For intermediate limb darkening we find:

$$\frac{l_{min}}{l_{max}} = 1 - \frac{15 + x}{15 - 5x}(1 + y)\,\epsilon = 1 - N\epsilon$$

In general the intensity in the maximum is:

$$l = 1 - \frac{15 + x}{15 - 5x}(1 + y)\,\epsilon \sin^2 i \cos^2\theta \tag{353}$$

$$l = l_{45} - \frac{15 + x}{30 - 10x}(1 + y)\,\epsilon \sin^2 i \cos 2\theta$$

This formula can be condensed:

$$l = 1 - \tfrac{1}{2} Nz \cos^2\theta = l_{45} - \tfrac{1}{4} Nz \cos 2\theta \tag{354}$$

From the slope we find:

$$\tfrac{1}{2} Nz = \frac{15 + x}{15 - 5x}(1 + y)\,\epsilon \sin^2 i \tag{355}$$

The first factor runs between 1.0 and 1.6, the second factor between 1.06 and 2.25; the product between 1.06 and 3.60. As stated above the interval is smaller in practice. The best values for the product are $N = 2.2$ for $x = 0.4$; $N = 2.6$ for $x = 0.6$; $N = 3.2$ for $x = 0.8$; $N = 3.6$ for $x = 1.0$. The determination of the constants happens again in the same way. The rectification proceeds as described earlier. In the dynamical condition we again have the correction factor $E(\theta) = 1 - Nz \cos^2\theta$; the other method also follows as before.

117. *Reflection effect.* Each component receives light from the other star that in turn is reflected towards us. We shall first consider the very simple case in which the star disks reflect this light like mirrors. The subscripts c and h will indicate respectively the cooler and hotter star. Let L_c and L_h be the intrinsic luminosities of the stars as seen on the non-illuminated hemispheres. We shall call $2s_c$ and $2s_h$ the respective total reflections. Further, we introduce l_c and l_h respectively as the mean intensity for each star. The two pairs of hemispheres then show the following intensities (Figure 158):

$L_c = l_c - s_c; L_c + 2s_c = l_c + s_c; L_h + 2s_h = l_h + s_h; L_h = l_h - s_h$

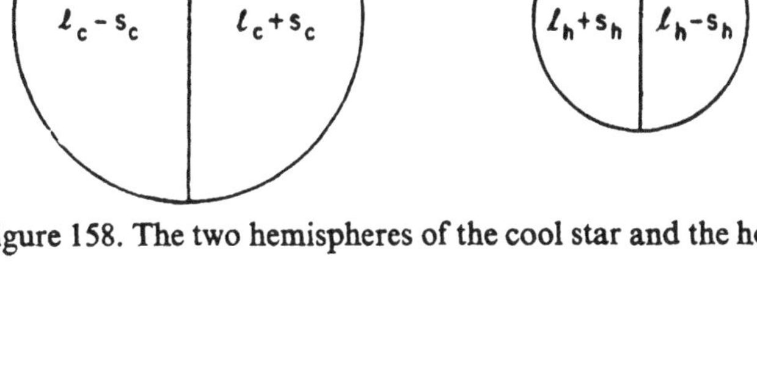

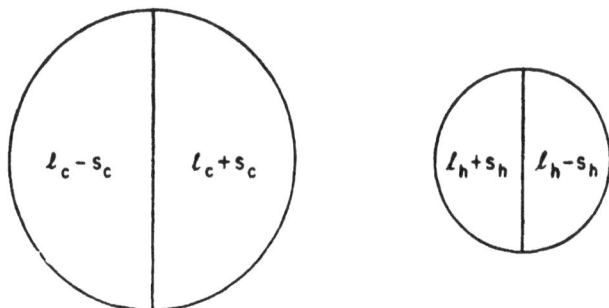

Figure 158. The two hemispheres of the cool star and the hot star.

During primary minimum the star with the greater surface luminosity I_h or the hotter star is eclipsed; during secondary minimum the hotter star is in front. We then have:

$$\frac{L_h}{L_c} = \frac{r_h^2 I_h}{r_c^2 I_c} = k^2 \frac{I_h}{I_c} \tag{356}$$

The reflection, however, is proportional to both the intensity of the other radiating star and to the area of the reflecting component.

$$\frac{s_c}{s_h} = \frac{L_h r_c^2}{L_c r_h^2} = \frac{I_h}{I_c} = \frac{1 - l_1}{1 - l_2} \tag{357}$$

The reflection from the cool star is therefore proportional to the surface brightness of the hot star and is thus the larger; $s_c - s_h$ is positive. The ratio of the reflections can be found from the ratio of the surface brightnesses, which for uniform light can be found from the ratio of the depths of the minima after a provisional rectification has been made. In case of limb darkening we can start in the same way and after the provisional solution of the orbital elements has been made we can compute the ratio of the mean surface luminosities. If necessary we can repeat the whole computation with this better value. Just outside primary minimum we again see the hotter star which emits the smaller reflection. In addition, we see the intrinsic light of the cool star. Around $\theta = 0°$ therefore we find the minimum reflection effect.

Let us first work with the averaged light. For a given phase angle, φ or θ, we can write down the observed light; between the stars there is a phase difference of 180° at any time:

$$\left.\begin{aligned} l_h + s_h \cos\varphi &= L_h + s_h + s_h \cos\varphi = L_h + s_h(1 + \cos\varphi) \\ l_c - s_c \cos\varphi &= L_c + s_c - s_c \cos\varphi = L_c + s_c(1 - \cos\varphi) \end{aligned}\right\}$$

We can write this as follows:

$$\left.\begin{aligned} L_h + 2s_h f(\varphi) &= L_h + 2s_h(0.5 + 0.5\cos\varphi) \\ L_c + 2s_c f(\varphi') &= L_c + 2s_c(0.5 - 0.5\cos\varphi) \end{aligned}\right\} \tag{358}$$

The value of the phase function $f(\varphi) = \frac{1}{2}(1 + \cos\varphi)$ runs between 0 and 1 for the possible values of φ.

The effect of reflection on the intensity of the maxima is:

$$l = (l_c + l_h) - (s_c - s_h)\cos\varphi$$
$$l = (L_c + L_h) + (s_c + s_h) - (s_c - s_h)\cos\varphi \qquad (359)$$

For $\varphi = 90°$ we have $\cos\varphi = 0$ and we can write:

$$l_{av} = l_{90} = l_c + l_h, \qquad l_* = L_c + L_h$$

This intrinsic light l_* cannot any longer be taken equal to unity. For the variation of the light in the maximum we can write (Figure 159):

$$l = l_{90} - (s_c - s_h)\sin i\cos\theta \qquad (360)$$
$$l = l_* + (s_c + s_h) - (s_c - s_h)\sin i\cos\theta \qquad (361)$$

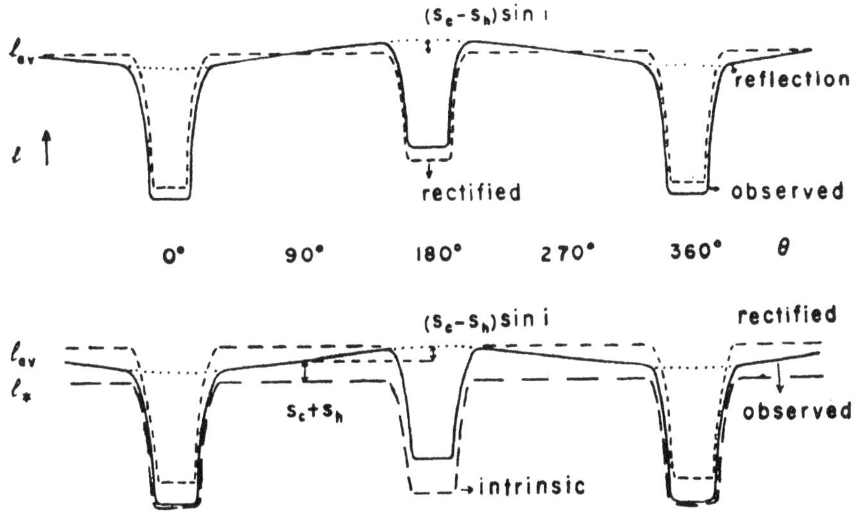

Figure 159. Influence of the reflection effect on the observed lightcurve. In addition the intrinsic lightcurve (without reflection) and the rectified lightcurve are given.

From the slope of the plot between l and $\cos\theta$ we find $(s_c - s_h)$ $\sin i$, and if we set $i = 90°$, the difference $(s_c - s_h)$ follows. With the difference and the ratio known we can find s_c and s_h separately. From a provisional solution of the orbital elements a better value of i can be inserted. The term $(s_c + s_h)$ in intensity does not give a constant in magnitudes. Failure to apply this term distorts the branches of the light curve somewhat, and we will find an incorrect k. If the observations are not accurate, it does not affect the determination appreciably.

For identical stars $s_c - s_h = 0$, the $\cos\theta$ term vanishes; the minima have the same depths. We know $s_c = s_h$ but cannot find the actual numerical value in this case. This may happen for example with W Ursae Majoris stars, which will certainly have a reflection effect because the stars are so close in space. On the other hand an Algol type star may show only a small $\cos\theta$ term in the maximum, though the depths of minima can be very unequal. These stars are so far from each other that the reflections are small and consequently their difference smaller still. The fact that the ratio may be large is then of no importance.

118. *Approximate rectification.* We can now write the observed intensity in the maximum as follows, where we shall take all constants as positive values. This is contrary to the literature, where only positive signs are used but the constants have negative values sometimes (Figure 160).

$$l = B_0 - B_1 \cos\theta - B_2 \cos^2\theta = A_0 - A_1 \cos\theta - A_2 \cos 2\theta \quad (362)$$

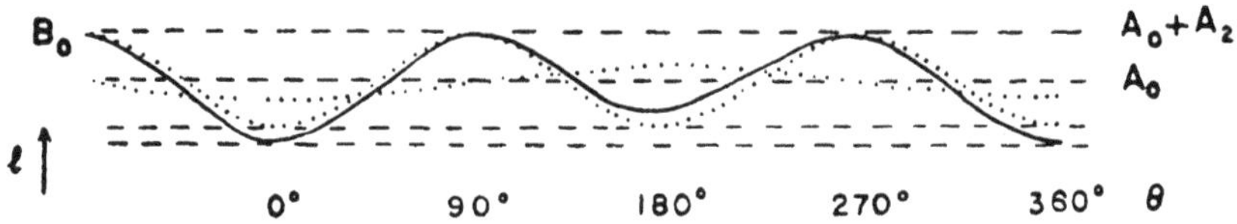

Figure 160. The observed light variation in the maxima as caused by oblateness and reflection.

The maximum values at $\theta = 90°$ and $\theta = 270°$ are the same. We can determine these constants from the observed maxima as follows. Take as the three intensities: l_a (θ and $360° - \theta$), l_b ($180° - \theta$ and $180° + \theta$), l_c ($90°$ and $270°$) and we find:

$$\left.\begin{aligned} l_a &= B_0 - B_1 \cos\theta - B_2 \cos^2\theta = A_0 - A_1 \cos\theta - A_2 \cos 2\theta \\ l_b &= B_0 + B_1 \cos\theta - B_2 \cos^2\theta = A_0 + A_1 \cos\theta - A_2 \cos 2\theta \\ l_c &= B_0 \qquad\qquad\qquad\qquad\quad\; = A_0 + A_2 \end{aligned}\right\}$$

From the third equation we find B_0.

$$\tfrac{1}{2}(l_b - l_a) = B_1 \cos\theta = A_1 \cos\theta, \quad \text{gives} \quad B_1 = A_1$$

$$\begin{aligned} l_a = l_c - \tfrac{1}{2}(l_b - l_a) - B_2 \cos_2\theta &= l_c - A_2 - \tfrac{1}{2}(l_b - l_a) - A_2 \cos 2\theta \\ &= l_c - \tfrac{1}{2}(l_b - l_a) - 2A_2 \cos^2\theta \end{aligned}$$

Thus $B_2 = 2A_2$ is known.

Naturally the constants are best determined by least squares or Fourier analysis, using all the normal points in the maximum. The figure shows the meaning of the constants. Here we consider the $\cos\theta$ term caused by reflection and $\cos^2\theta$ or $\cos 2\theta$ only by oblateness. For rectification divide the observed intensity for each normal point by the value on the curve drawn through these normal points corresponding to the same θ.

$$l_r = \frac{l}{B_0 - B_1 \cos\theta - B_2 \cos^2\theta} = \frac{l}{A_0 - A_1 \cos\theta - A_2 \cos 2\theta} \tag{363}$$

Here l_r is the rectified value of the normal point. After rectification we find for the mean value of all normal points in the maxima: $\overline{l}_{r,\,max} = 1$; thus we have flat maxima. The rectified light curve can then be analyzed as before. For visual or photographic light curves where the accuracy is not extremely high this rectification is sufficient.

119. *Reflection and reradiation.* Unfortunately the stars are not such perfect mirrors as we have assumed. A somewhat better approximation would be to consider all the surface elements of

the sphere or ellipsoid as mirrors and study the combined reflection. But after receiving light from the other radiating star the reflecting component partly absorbs it and then reradiates it in another wavelength. This is an extremely complicated process; an entire rigorous solution still has not been completed but the theory is adequate for many actual cases. For example the sizes of the components have been taken into account up to $r = 0.20$ which is sufficient for most eclipsing stars but not for W Ursae Majoris stars. For these last stars additional complications arise by the fact that often they are surrounded by gas streams and shells, for which it is difficult to apply a correction.

Besides the term in $\cos\theta$ already found, we shall find a new constant term and a term in $\cos^2\theta$ or $\cos 2\theta$. Here we shall first consider the $\cos^2\theta$ term, since then the coefficients become simple and symmetrical. This process also shows the relation of added terms with the previous simple picture.

For the hot star the so-called phase function is theoretically found to be of the form:

$$f(\varphi) = 0.2 + 0.4\cos\varphi + 0.2\cos^2\varphi \qquad (364)$$

The phase function runs between 0 and 0.8 for the possible values of φ. Note that the coefficients are symmetrical.

Let L_c and L_h again be the intrinsic luminosities. Recalling that the phases of both stars differ by 180° for any given time, we can write down the observed intensity for both the hot and the cool stars:

$$\left.\begin{aligned} L_h + L_c r_h^2 f(\varphi) &= L_h + L_c r_h^2 (0.2 + 0.4\cos\varphi + 0.2\cos^2\varphi) \\ L_c + L_h r_c^2 f(\varphi) &= L_c + L_h r_c^2 (0.2 - 0.4\cos\varphi + 0.2\cos^2\varphi) \end{aligned}\right\}$$

Comparing the coefficients of $\cos\varphi$ found here with those used earlier in the simple picture, we can call:

$$s_h = 0.4\,L_c r_h^2\,, \qquad s_c = 0.4\,L_h r_c^2 \qquad (365)$$

We can rewrite the intensities of the hot and cool stars as we observe them for a given time:

$$\left.\begin{aligned} L_h + L_c r_h^2 f(\varphi) &= L_h + s_h(\tfrac{1}{2} + \cos\varphi + \tfrac{1}{2}\cos^2\varphi) \\ L_c + L_h r_c^2 f(\varphi) &= L_c + s_c(\tfrac{1}{2} - \cos\varphi + \tfrac{1}{2}\cos^2\varphi) \end{aligned}\right\}$$

In the maximum the total light received is now:

$$l = (L_c + L_h) + \tfrac{1}{2}(s_c + s_h) - (s_c - s_h)\cos\varphi + \tfrac{1}{2}(s_c + s_h)\cos^2\varphi \tag{366}$$

Introducing the intrinsic light l_* and θ, we have (Figure 161):

$$l = l_* + \tfrac{1}{2}(s_c + s_h) - (s_c - s_h)\sin i\cos\theta + \tfrac{1}{2}(s_c + s_h)\sin^2 i\cos^2\theta \tag{367}$$

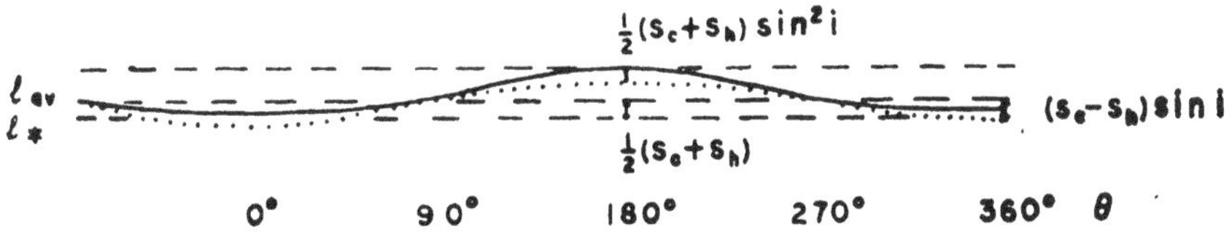

Figure 161. The effect of reflection and reradiation on the light intensity of the maxima.

Again $s_c - s_h$ is positive. For $\theta = 180°$ we find l_{max}, but in practice this occurs at the phase of secondary minimum. We have found in addition a constant term and an additional term in $\cos^2\theta$. This latter term combines with the oblateness, but the signs are opposite. The $\cos\theta$ term is caused only by reflection and can be determined directly from the observations. Both other terms have to be computed. All this holds for bolometric light. For monochromatic light the distribution of light in the spectra must be studied because reradiation takes place in another wavelength. For the late type spectra there are numerous absorption lines; for the early type spectra the Lyman and Balmer absorptions play a role. J. E. Merrill has presented a graph into which we can enter with the ratio s_c/s_h found from the ratio of the depths of minima as derived after the provisional rectification, and find a ratio S_c/S_h. In S_c and S_h the so-called luminosity efficiencies of the stars are taken into account on the basis of grey body theory. For monochromatic light we can replace s by S, and l_* by L_* without changing the structure of the formula.

120. *Correct rectification.* Let us collect all the considered effects that can cause a variation in the maximum light. The intrinsic light L_* varies then by oblateness.

$$l = L_* + \tfrac{1}{2}(S_c + S_h) - (S_c - S_h)\sin i \cos\theta + \tfrac{1}{2}(S_c + S_h)\sin^2 i\cos^2\theta \quad (368)$$

$$L_* = L\left\{1 - \frac{15 + x}{15 - 5x}(1 + y)\epsilon\right\}\sin^2 i\cos^2\theta$$

The rectification for reflection now proceeds as follows. We note that only the inner hemispheres are affected by eclipses. Therefore add enough luminosity to the outer face of each component to bring it up to equality with the illuminated sides. For the unilluminated side the phase is $\theta + 180°$ so that the cos θ term changes sign. We therefore add the amount:

$$\Delta l = \tfrac{1}{2}(S_c + S_h) + (S_c - S_h)\sin i\cos\theta + \tfrac{1}{2}(S_c + S_h)\sin^2 i\cos^2\theta \quad (369)$$

This addition of intensity has no equivalent in magnitudes. After rectification for reflection we have therefore:

$$l'_r = L + S_c + S_h + \left\{S_c + S_h - L\frac{15 + x}{15 - 5x}(1 + y)\,\epsilon\right\}\sin^2 i\cos^2\theta \quad (370)$$

Here l'_r is the rectified value of the normal point. Let l_θ be the value on the curve drawn through these normal points corresponding to the same θ. We have still to rectify for oblateness;

this we can do in the usual way. Divide the intensity of each normal point by l_θ.

$$l''_r = \frac{l'_r}{l_\theta}, \qquad \text{thus} \qquad \overline{l''_{r,\,max}} = 1 \tag{371}$$

After both rectifications the curve will again have two flat maxima.

In practice we often work as follows. The maximum values at $\theta = 90°$ and $270°$ are the same. The observed light variation in the maxima can be expressed as:

$$l = B_0 - B_1 \cos\theta - B_2 \cos^2\theta = L_* + D_0 - D_1 \cos\theta + D_2 \cos^2\theta \tag{372}$$

All the constants are taken positive. The first three constants are determined as before from the light curve. The last three which represent only the reflection effect are found as follows. We know $D_1 = B_1$ because the cos θ term contains only the reflection effect. Further, by comparison of the coefficients $D_2 = D_0 \sin^2 i$. We have thus to determine only D_0.

$$D_0 = \tfrac{1}{2}(S_c + S_h), \qquad D_1 = B_1 = (S_c - S_h)\sin i$$

$$D_0 = \tfrac{1}{2}\left(\frac{S_c + S_h}{S_c - S_h}\right)\frac{B_1}{\sin i} \tag{373}$$

We can first take $i = 90°$. From Merrill's graph we find S_c/S_h and by a property of ratios determine the above factor. In this way the D_0 and eventually D_2 can be computed. For the intrinsic light we have:

$$L_* = (B_0 - D_0) - (B_2 + D_2)\cos^2\theta$$

$$L_* = (B_0 - D_0)\left\{1 - \frac{B_2 + D_2}{B_0 - D_0}\cos^2\theta\right\} \tag{374}$$

The intrinsic light is affected only by oblateness, and consequently we expect the form $1 - \tfrac{1}{2}Nz\cos^2\theta$. The eclipses are affected by reflection because the illuminated faces are eclipsed.

For rectification of reflection we have to add luminosity to the unilluminated hemispheres. Because of the phase difference of 180° the $D_1 \cos\theta$ takes a positive sign.

$$l'_r = l + D_0 + D_1\cos\theta + D_2\cos^2\theta = (B_0 + D_0) - (B_2 - D_2)\cos^2\theta \tag{375}$$

The rectification for oblateness is:

$$l''_r = \frac{l'_r}{(B_0 + D_0) - (B_2 - D_2)\cos^2\theta}, \quad \overline{l''}_{r,\,\max} = 1 \tag{376}$$

After rectification the mean value of the normal points in the maxima is unity; outside eclipse we now find flat maxima.

Russell and Merrill use the cos 2θ term. The coefficients lose their symmetry and the procedure is somewhat more complicated. The phase function becomes:

$$\begin{aligned} f(\varphi) &= 0.3 + 0.4\cos\varphi + 0.1\cos 2\varphi \\ f(\theta) &= 0.2 + 0.4\sin i\cos\theta + 0.2\sin^2 i\cos^2\theta \\ &= 0.2 + 0.4\sin i\cos\theta + 0.1\sin^2 i\,(1 + \cos 2\theta) \\ &= (0.2 + 0.1\sin^2 i) + 0.4\sin i\cos\theta + 0.1\sin^2 i\cos 2\theta \\ &= (0.3 - 0.1\cos^2 i) + 0.4\sin i\cos\theta + 0.1\sin^2 i\cos 2\theta \end{aligned}$$

The intensity of light in the maxima becomes:

$$l = l_* + \frac{1}{4}(s_c + s_h)(2 + \sin^2 i) - (s_c - s_h)\sin i\cos\theta + \frac{1}{4}(s_c + s_h)\sin^2 i\cos 2\theta$$

For monochromatic light we again replace s by S, and l_* by L_*. The observed light variation in the maxima can be written as:

$$l = A_0 - A_1\cos\theta - A_2\cos 2\theta = L_* + C_0 - C_1\cos\theta + C_2\cos 2\theta$$

The first three constants are determined from the light curve as before. The last three are found as follows. We know $C_1 = A_1$; for $i = 90°$ we see that $3C_2 = C_0$. We have to determine only C_0.

$$C_0 = \frac{1}{4}(S_c + S_h)(2 + \sin^2 i) = \frac{1}{4}\left(\frac{2}{\sin i} + \sin i\right)\frac{S_c + S_h}{S_c - S_h}A_1$$

$$= \frac{1}{4}(3 - \cos^2 i)\frac{S_c + S_h}{S_c - S_h}A_1 \operatorname{cosec} i.$$

For $i = 90°$ we get:

$$C_0 = \frac{3}{4}\frac{S_c + S_h}{S_c - S_h}A_1$$

For the intrinsic light we have:

$$L_* = (A_0 - C_0) - (A_2 + C_2)\cos 2\theta$$
$$= (A_0 - C_0 + A_2 + C_2) - 2(A_2 + C_2)\cos^2\theta$$
$$= (A_0 - C_0 + A_2 + C_2)\left\{1 - \frac{2(A_2 + C_2)}{A_0 - C_0 + A_2 + C_2}\cos^2\theta\right\}$$

The rectification in reflection becomes:

$$l'_r = l + C_0 + C_1\cos\theta + C_2\cos 2\theta = (A_0 + C_0) - (A_2 - C_2)\cos 2\theta$$

For the rectification in oblateness we get:

$$l''_r = \frac{l'_r}{(A_0 + C_0) - (A_2 - C_2)\cos 2\theta}, \qquad \overline{l''}_{r,\,max} = 1$$

Again we find flat maxima after rectification.

Sometimes the constant and terms in cos θ, cos²θ or cos 2θ, are not sufficient to describe the maximum light, namely when the maximum values at θ = 90° and θ = 270° are unequal. We are forced to introduce sin θ and sin 2θ terms. We have no simple theoretical explanation for them, but we can call them perturbations. These may be caused by the fact that our model is only a certain approximation of the real star. Consider for example the three unequal axes of the real star. In this case both minima will be affected too. The oblateness of both stars may be different in which case only one minimum will be affected. Another possibility is that an overluminous region is situated unsymmetrically on one of the stars, so that for example the region would be visible at θ = 90° but invisible at θ = 270°.

Again one minimum will be affected, which will become unsymmetrical and therefore apparently shifted in time.

The light in the maxima is then expressed in a Fourier series. J. E. Merrill has given a simple method of determining the constants. It can also be done by least squares or Fourier analysis. The amplitude of the sin θ term depends on the difference of the maximum intensities in $\theta = 90°$ and $\theta = 270°$. We rectify first for the sin θ and sin 2θ terms after which the minima should become symmetrical. After that, the rectification proceeds as described before. But it is well to remember that the orbital elements found are somewhat artificial in this case.

121. *Visible and invisible companions.* It occasionally so happens, that the light of an additional star, having a very small angular distance to the eclipsing star, has been included in the photographic or photo-electric measurements. This companion may be an optical or a physical one. If the light of the companion is visible, one can correct in the following way. Measure visually the magnitude difference Δm between this component and the maximum of the eclipsing variable.

In intensities we have for the observed maximum light:

$$(L_s + L_g) + L_c = 1 \tag{377}$$

From the magnitude difference we find now the light of the component L_c in terms of the maximum light $(L_s + L_g)$ of the eclipsing system as a unit.

$$\gamma = \frac{L_c}{L_s + L_g}, \qquad L_c = \gamma(L_s + L_g) \tag{378}$$

We free the observed light now from the light of the companion; that is we subtract L_c from the maximum light and from the observed light at any phase angle. For the maximum we get:

$$L_s + L_g = 1 - L_c = 1 - \gamma(L_s + L_g)$$

$$L_s + \mathrm{L}_g = \frac{1}{1 + \gamma} < 1 \tag{379}$$

In intensity the light curve is just shifted but keeps the same form; in magnitude the light curve is shifted and changed in shape.

For orbital computation we were allowed to multiply the intensity by a factor or add the corresponding term to the magnitudes. We multiply all the intensities now by the factor $(1 + \gamma)$ so that the maximum intensities become unity again:

$$L'_s + L'_g = (1 + \gamma)(L_s + L_g) = 1 \tag{380}$$

Compared with the observed light we see now that the depth of the minimum becomes larger and the slope of the branches different. Corresponding changes occur in the magnitudes, and thus in the light curve. In other words, not making this correction will give the wrong orbital elements.

In case the component is invisible or if a gas stream surrounds the system the orbital elements derived from the observed light curve will obviously be unsatisfactory. In such a case it may not be possible to find a solution which gives a satisfactory fit to the light curve. If such a component or gas stream is suspected one can assume a Δm and see whether the light curve freed from this light gives a better orbital solution. After some trial and error one may succeed in this and at the same time find thus an estimate of the Δm.

122. ***Résumé.*** For an accurate light curve the rectification for reflection and oblateness must be done separately. The rectified light curve has two flat maxima. The depths of minima give a relation between α_0 and k. The branches give x, k, r_g, i for complete eclipses; for partial eclipses we assume x and find a shape relation between α_0 and k; then α_0, k, r_g, i follow. Returning to the maximum, the coefficient of $\cos^2\theta$ or $\cos 2\theta$ gives us certain information; x is already known, y can be found theoretically; thus the oblateness follows. The coefficient of $\cos\theta$ gives the

relative reflection effect; the ratio of the reflections for both components can be found for uniform light from the ratio of the depths of minima. The amplitude of the $\cos^2\theta$ or $\cos 2\theta$ terms becomes larger the greater the limb darkening (1.0 to 1.6); the factor involving the gravitation makes it still larger by a factor of about 2, but the reflection effect in this term opposes the gravitation effect and decreases the amplitude again. The oblateness is of the order of 0.1 at most.

123. *Eccentricity of the orbit.* The case is now: elliptical orbit, spherical stars. Now the covered areas need not be the same during eclipses. In addition we may have the combination of one complete eclipse and one partial one. Thus the following simple formulae for uniform light no longer hold in the general case and are only true for some favored cases:

$$k^2 = \frac{1 - l_a}{l_t}, \qquad a_0 = C + \frac{D}{k^2}, \qquad \frac{I_h}{I_c} = \frac{1 - l_1}{1 - l_2}$$

It is clear, therefore, that the process of determining the elements is more difficult. We shall first consider two special orientations and then the general case, all for $i = 90°$.

(1) Major axis in the plane of the sky. According to Kepler's second law the areas of the orbit on either side of the line of sight are not swept in equal times. Thus the minima are not separated by $\frac{1}{2}P$, but they do have equal durations. Let us call D the phase difference, where the phase itself runs from 0 to 1; then the phase shift is $D - \frac{1}{2}$. For a small eccentricity the second law gives (Figure 162):

$$\frac{D}{1} = \frac{\frac{1}{2}\pi\, ab \pm 2b\, ae}{\pi\, ab} = \frac{1}{2} + \frac{2e}{\pi}$$

If we call t the duration of the eclipse we then find:

$$e = \pm\frac{\pi}{2}(D - \tfrac{1}{2}), \qquad t_1 = t_2 \qquad (381)$$

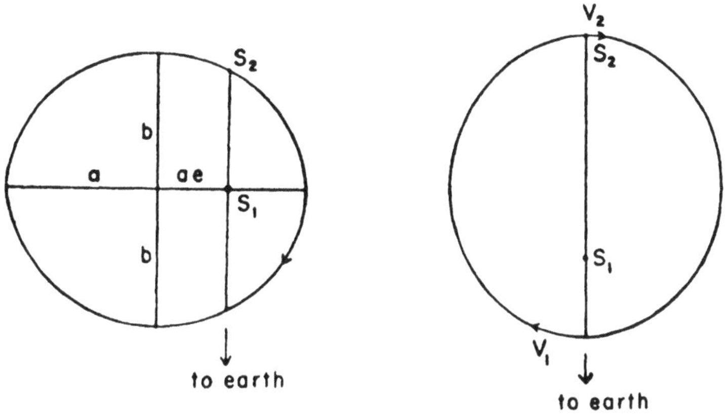

Figure 162. Left: Major axis in the plane of the sky gives a phase shift. Right: Major axis in the line of sight gives unequal durations of the minima.

(2) Major axis in the line of sight. The minima are now exactly $\frac{1}{2}P$ apart, just as in the case of a circular orbit. However, the durations of eclipses are now quite different, while for a circular orbit they are the same. When the sizes of the stars are given, the durations of eclipses depend only on the velocities of the stars. Let the duration at apastron be t_1 and at periastron t_2; then $t_1 > t_2$. If V is the speed, according to the second law we have:

$$(a + ae)\,V_1 = (a - ae)V_2, \qquad \frac{1+e}{1-e} = \frac{V_2}{V_1} = \frac{t_1}{t_2}$$

A property of the ratios gives us then:

$$e = \frac{t_1 - t_2}{t_1 + t_2}, \qquad D - \tfrac{1}{2} = 0 \qquad (382)$$

(3) Now consider the case for $i = 90°$ in which the orientation of the major axis is in some general direction. We will then find both a phase shift and unequal durations of the minima. Draw a

line through the center parallel to the line of sight. According to Kepler's second law we have (Figure 163):

$$\frac{D}{1} = \frac{\frac{1}{2}\pi ab + 2b\, ae\cos\omega}{\pi ab} = \tfrac{1}{2} \pm \frac{2e\cos\omega}{\pi}$$

Thus for the phase shift we find:

$$e\cos\omega = \pm\frac{\pi}{2}(D - \tfrac{1}{2}) \qquad (383)$$

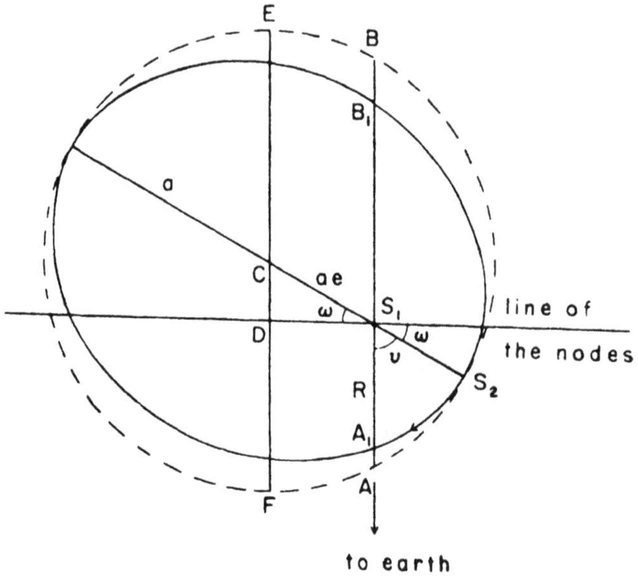

Figure 163. For the general case one finds a phase shift and unequal durations of the minima.

If we compute this more accurately, we find $(e^3/3\pi)\cos 3\omega$ for the next term so that we have only odd powers. We can also take into account the inclination and find the general result:

$$e\cos\omega = \pm\frac{\pi(D - \frac{1}{2})}{1 + \operatorname{cosec}^2 i} \qquad (384)$$

Along the line of sight through the center we have:

$$\frac{DF}{DE} = \frac{a + ae \sin \omega}{a - ae \sin \omega} = \frac{1 + e \sin \omega}{1 - e \sin \omega}$$

This ratio is approximately the same as the ratios S_1A/S_1B and S_1A_1/S_1B_1; this latter ratio remains the same when projected through $90° - i$.

$$\frac{1 + e \sin \omega}{1 - e \sin \omega} = \frac{t_1}{t_2}$$

From this we find the following formula:

$$e \sin \omega = \frac{t_1 - t_2}{t_1 + t_2} \tag{385}$$

From both formulae we can solve for e and ω, but the latter formula is only approximate. However, the duration of eclipse usually cannot be determined very accurately in practice because the times of external tangency are difficult to fix. Perhaps the best thing to do in most cases is to take e from spectroscopic data and find ω from $e \cos \omega$.

The dynamical condition also changes somewhat. During eclipse the distance between the stars is given by the formula:

$$R = \frac{a(1 - e^2)}{1 + e \cos v} = \frac{1 - e^2}{1 + e \cos v} \tag{386}$$

Here we have taken the semi-major axis of the orbit equal to unity, but we cannot set the radius vector R equal to unity. For primary mid-eclipse we have $v + \omega = 90°$, and since ω is known, we have the true anomaly v. With help of the derived e we find R in terms of semi-major axis as a unit. Let $\theta' = v + \omega - 90°$ corresponding to a circular orbit. For a given e the v is tabulated by F. Schlesinger and S. Udick. The θ' can be found for any normal point; for mid-eclipse $\theta' = 0$. The dynamical condition becomes:

$$\delta^2 = R^2 \{\cos^2 i + \sin^2 i \sin^2 \theta'\} \tag{387}$$

The duration of eclipse is short compared with the orbital period. During eclipse, therefore, we can consider the orbit a circle of radius R equal to the computed radius vector.

$$\left(\frac{\delta}{r_g}\right)^2 = R^2\left\{\frac{\cos^2 i}{r_g^2} + \frac{\sin^2 i}{r_g^2}\sin^2\theta'\right\}$$

$$\left(\frac{\delta}{r_g}\right)^2 \cdot \frac{1}{R^2} = \frac{\cos^2 i}{r_g^2} + \frac{\sin^2 i}{r_g^2}\sin^2\theta' \tag{388}$$

The dynamical condition is now of the form:

$$\frac{u}{R^2} = A + B\sin^2\theta' \tag{389}$$

Thus we replace u by u/R^2 and θ by θ' as compared with the formula for a circular orbit. For oblate stars we get the additional correction factor $E(\theta')$. The maxima are again curved, but the picture is somewhat more complicated.

124. *Rotation of the line of apsides.* Some eclipsing stars do not have a strictly constant period. This variation may be caused by a third body. The two components that produce the eclipse then have a gravity center, which revolves around the common gravity center of the three components. There is involved a so-called stellar light time which can be observed as a variation of the period. The times of primary and secondary minima are affected in the same way.

Another case is the rotation of the major axis or the line of apsides, observed in about twenty stars. This period is always very long in comparison with the normal period of revolution. There is a similar situation in our planetary system.

A spherical star causes a gravitational field proportional to $1/d^2$ when d is the distance between the center of the star and an outside point. An oblate star has a gravitational field with, in addition, terms in $1/d^3$, $1/d^4$ etc. This distortion of the field causes a perturbation in the elliptical orbit in such a way that the semi-major axis begins to rotate.

In Figure 164 we have first sketched the minima for the major axis in the line of sight; they are then half a period apart. As soon as the major axis starts to rotate, we will find a phase

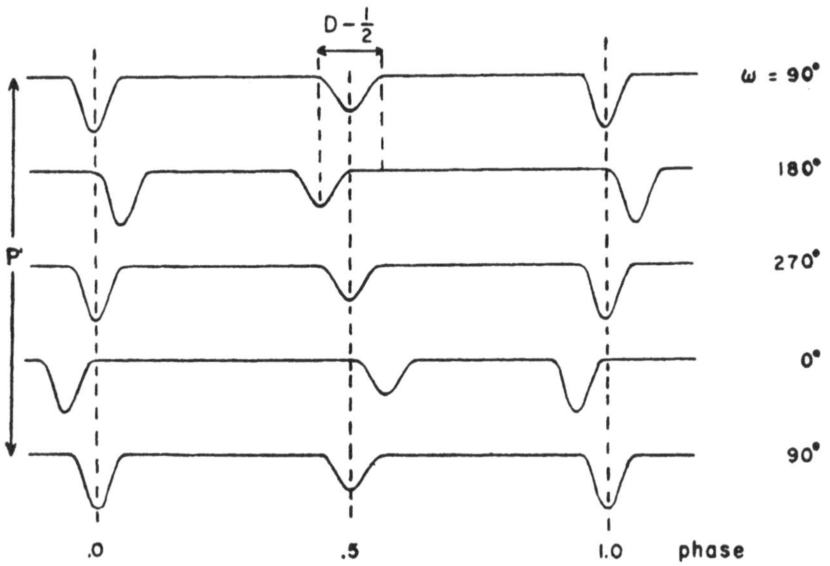

Figure 164. Phase shift caused by rotation of the line of apsides.

shift. Horizontally we have laid off in days the normal period P; vertically we have successively drawn the light curves as they change throughout the much longer period P'. The primary minimum has a periodic change around the strictly linear ephemeris and the secondary minimum shows the opposite change. The periodic change can better be studied if we plot $O - C$ in days against the number of periods, showing the secondary term in the period (Figure 165). For $i = 90°$ and for direct motion we now find the relation theoretically:

$$O - C = A \sin\left(\omega - \frac{\pi}{2}\right) - B \sin 2\left(\omega - \frac{\pi}{2}\right) \qquad (390)$$

$$A = \frac{e}{\pi} = 0.32e, \qquad B = \frac{3e^2}{8\pi} = 0.12\, e^2$$

For the secondary minimum, the A has a negative sign in front. The e is constant; the ω has a period P'.

The formula includes a sine of the argument $(\omega - \pi/2)$ and also the sine of twice the argument. Both curves intersect on the horizontal axis. The points where $O - C = 0$ correspond to those times when the major axis is in the line of sight, that is when $\omega = \pi/2$ and $\omega = 3\pi/2$. For each point on the observed curve the ω can be found. Then we can find the constants A and B as half the amplitudes of the respective sine

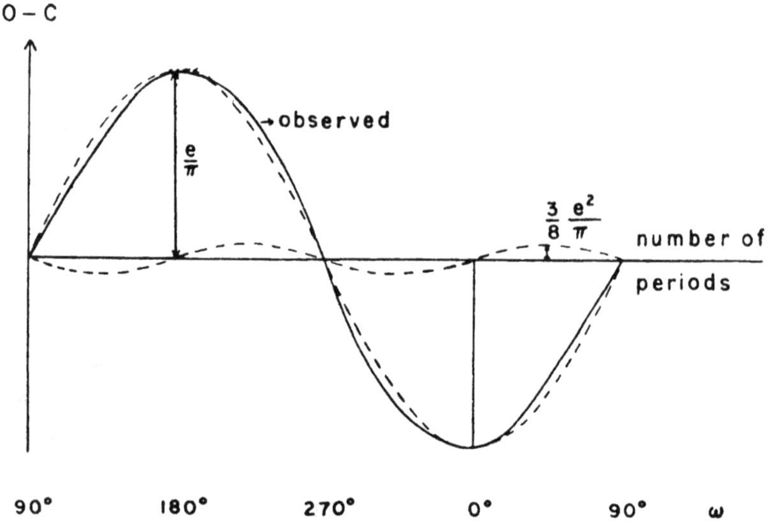

Figure 165. The periodic change of the time of primary minimum using a linear period, as caused by the rotation of the line of apsides.

waves. The first constant is found at $\omega = \pi$; the second one at $\omega = 3\pi/4$. This gives us the e.

In practice P' is often long and then only part of Figure 165 has been observed. We determine the e according to paragraph 123 and substitute this value in formula (390). The observed values of $O - C$ give us the ω's and thus the period P' expressed in P.

The components of an Algol star are so far away from each other in space that the tidal action is small. Only axial rotation can produce the oblateness necessary for the rotation of the line of apsides. The areas observed are constant during the period, so that the maxima are flat. The rotation can be discovered since the absorption lines in the spectrum become broader. For rotation alone, H. N. Russell found:

$$\frac{P}{P'} = C_1\left(\frac{r_1}{R}\right)^5\left(1 + \frac{\mathfrak{M}_2}{\mathfrak{M}_1}\right) + [\text{star } 2] \qquad (391)$$

For star 2 we have similar expression with the subscripts 1 and 2 interchanged. For β Lyrae stars we have mainly tidal distortion (axes a, b, b), and here Russell found:

$$\frac{P}{P'} = C_1\left(\frac{r_1}{R}\right)^5\left(1 + 7\frac{\mathfrak{M}_2}{\mathfrak{M}_1}\right) + [\text{star } 2] \qquad (392)$$

The coefficient of the mass ratio is 1 for rotation alone, 6 for tidal distortion alone, together 7. There is no e in the formula. If we place the stars twice as far away from each other, then in the fifth power we introduce a factor 1/32 and at the same time increase P by a factor of 2.8 according to Kepler's third law. Because these are on opposite sides of the equality the total factor involved is 90. The period of apsidal rotation, P', quickly becomes so long that it cannot be observed. For the very short period eclipsing stars we could expect a short P' too, but here the orbits are practically circular, and we cannot talk about rotation of the semi-major axis. Thus only stars of intermediate period show this effect, and it has not been detected in all of them.

In the derivation of these formulae the assumption was made that the stars approximate ellipsoids whose major axes are along the line joining the centers of the two stars. The stars are thus considered as rigid during the motion having constant tides in an orbit of small eccentricity.

T. G. Cowling takes into account the variation of the tidal distortion with the mutual distance of the stars in the orbit. Further he assumes that the major axes of the stars remain in the line of the centers. His assumptions are fewer and probably more sound. Therefore his formula is customarily used:

$$\frac{P}{P'} = C_1\left(\frac{r_1}{R}\right)^5\left(1 + 16\frac{\mathfrak{M}_2}{\mathfrak{M}_1}\right) + [\text{star } 2] \qquad (393)$$

This formula is the limiting case for $e = 0$ in a more elaborate expression. We have seen already that for a circular motion the line of apsides is indeterminate. The above formula is therefore an approximation for orbits with small eccentricity. The constant C_1 can be computed by assuming a particular model. For the MacLaurin model of a homogeneous star $C_1 = 3/4$; for the Roche model where all the mass is supposed to be in the center $C_1 = 0$. For an actual star P and P' can be observed; r_1 in terms of R is found from the light curve; the mass ratio is found from the radial velocity curves or from the magnitude difference Δm assuming that the mass-luminosity relation holds. We can then find the constant C_1 which is always very small. Thus we find information about the density gradient of the star. The Roche model approximates the actual star.

125. *Position angle of the node*. Usually Ω cannot be determined, but when the components are of very early type, according to

theory the star light is polarized along the radius. The star as a whole shows no polarization because it is cancelled by symmetry, if we see the whole disk. But during an eclipse we may see only a crescent of the hot star. This gives a polarization component. By measuring the plane of polarization we could identify the plane of the orbit in space and thus find Ω.

When attempts were made to observe this, another effect was immediately found which was even more interesting. In the Milky Way each distant star shows constant polarization that differs from star to star. It is caused by the interstellar dust particles or grains through which the star light passed. In general the effect is more appreciable for more distant stars. The dust particles are being lined up with respect to the Milky Way plane by the general magnetic field of our galaxy.

126. *Extended atmospheres.* Until now we have considered geometrical eclipses. The edges of the stars were sharp, just as is the sun's edge if we forget about its prominences. However, red dwarf stars have prominence activity which has a high percentage effect on the intensity. They are observed as flare stars, which show a huge intensity increase in a short time, after which the normal intensity is restored. Also there will be star spots larger than sun spots. These phenomena are actually observed in the light curves of some red dwarf eclipsing stars and one should be cautious in using the theory of geometrical eclipses.

Another case where we have no geometrical eclipses is the following. Some supergiant stars are known to have an extended atmosphere without a sharp edge. The density of the stars is very low, something like a vacuum on earth. There are a few known eclipsing variables of this type, each with a normal component such as a *B* type. However, the small component does not show a real eclipse but becomes fainter and redder during ingress. It shows the same effect as the sun setting in a foggy atmosphere on the earth. We call this an atmospheric eclipse, instead of the body eclipse previously considered. The absorption is caused by the particles of the extended atmosphere in the line of sight. If f is the fraction of light of the smaller star which has been observed, the following formula holds:

$$1 - f = e^{-\tau} \tag{394}$$

Here τ is the so-called optical depth generally used in the theory of stellar atmospheres. There is not much difference between the shape of the branches for an atmospheric and a body eclipse. For a body eclipse the depths of the minima depend on the wavelength used, but the eclipse occurs simultaneously in any color. However, in violet light the atmospheric eclipse occurs earlier than in red light (Figure 166). The fraction of light lost at any given time is strongly dependent on the effective wavelength. If the smaller star is considered a point source compared

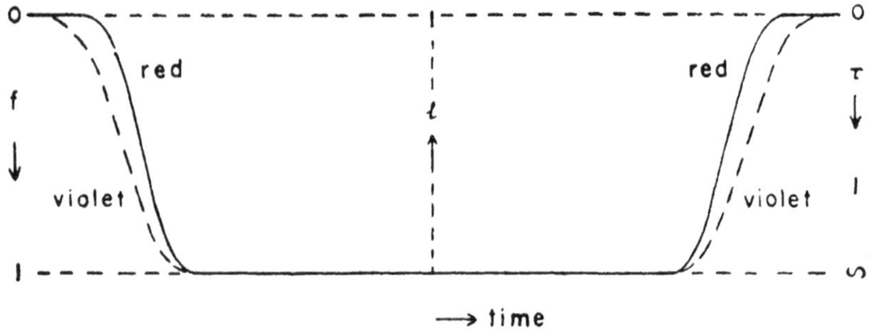

Figure 166. The atmospheric eclipse starts earlier and finishes later in violet light when compared with red light.

to the supergiant it is found that log τ must be proportional to the time. This plot demonstrates more clearly the small difference between both kinds of eclipses in any wavelength, because we can now compare a linear relation with a curved one. The largest deviation occurs near ingress or egress (see Figure 167).

Because the size of a supergiant is very large, the distance between the components is large and the period of revolution is long. The duration of eclipse may last several months; the branches or partial phases cover a period of several days or even weeks. Accurate light curves of these stars in different wavelengths are required but are difficult to obtain. Spectroscopic observations are of importance in order to learn something about successive layers of the atmosphere of the supergiant.

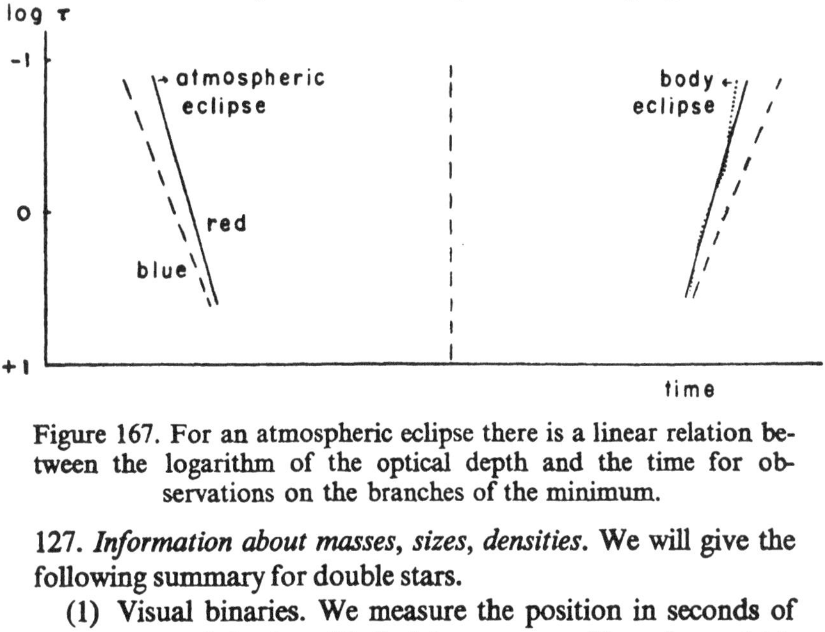

Figure 167. For an atmospheric eclipse there is a linear relation between the logarithm of the optical depth and the time for observations on the branches of the minimum.

127. *Information about masses, sizes, densities.* We will give the following summary for double stars.

(1) Visual binaries. We measure the position in seconds of arc, and the time. We find the apparent ellipse, from which we derive all the orbital elements. Only $\pm\, i$ can be found.

(a) Visual method. This gives the relative orbit. We find the semi-major axis of the relative orbit $a = a_1 + a_2$ in seconds of arc. If the parallax is known the a follows in astronomical units and the *sum of the masses* $\mathfrak{M}_1 + \mathfrak{M}_2$ expressed in solar masses.

(b) Photographic (with sequences), same as (1a).

(c) Astrometric. This gives the absolute orbit. We find the ratio of the major axes and the *mass ratio* $\mathfrak{M}_1/\mathfrak{M}_2 = a_2/a_1$.

(ac) Combination. This gives the *masses* $\mathfrak{M}_1$ and $\mathfrak{M}_2$ in solar masses.

(2) Spectroscopic binaries. We measure the radial velocity in km/sec and the time. We find the radial velocity curve.

(a) One curve. One finds $a_1 \sin i$ and the mass function $f(\mathfrak{M})$.

(b) Two curves. Here we can find the *mass ratio* $\mathfrak{M}_1/\mathfrak{M}_2 = K_2/K_1$. We derive $a \sin i$ in km, $\mathfrak{M}_1 \sin^3 i$, $\mathfrak{M}_2 \sin^3 i$, but do not find Ω. We can derive the ratio of the intensities.

(3) Eclipsing binaries. We measure the apparent magnitude and the time. We find the light curve from which we derive the i, the relative sizes and the relative intensities. We find neither Ω, nor the direction of motion.

(a) Algol type. Spherical stars, giving two nearly flat maxima.

(b) β Lyrae type. These are oblate stars, giving curved maxima.

(c) W Ursae Majoris type. These are oblate stars, which almost touch each other. We can find the mean density limit and the individual mean *densities* of each component.

Eclipsing variables give us information about limb darkening, oblateness, reflection, gravitation, rotation of the line of apsides, the *density gradient* (Roche model).

In practice the following combinations occur.

(2a) and (3) give us only the a_1 in kilometers.

(2b) and (3) give us the orbit and the radii in km, thus the *sizes*, and in addition the *masses* of the components. The individual *densities* follow.

(1c) and (2) give the inclination, thus the orientation in space.

(1c), (2) and (3) give complete orientation in space and all the information about the physical behavior of the stars.

The following holds for single stars, thus also for each component of a double star. By photographic methods we can find the proper motion, parallax, distance, absolute magnitude, and absolute intensity. We apply a correction to make this bolometric absolute intensity. Stefan's law gives the same per unit area.

$$I_{bol} = 4\pi r^2 \sigma T^4 = a\pi r^2 T^4 \tag{395}$$

We find the temperature T from the spectrum. The formula gives the area, the radius, thus the *size* and the volume. Find the mass from the mass-luminosity curve or from the above data. Mass = volume × density. Thus the mean *density* follows.

REFERENCES

GENERAL

H. N. Russell: *Pop. Astr.*, **54**, 162, 1946.
H. N. Russell and J. E. Merrill: *Contr. Princeton* No. 26, 1952.
L. Plaut: *Groningen Publ.* No. 54, 1950; No. 55, 1953.
Z. Kopal: *An introduction to the study of eclipsing variables.* Harvard Obs. Monograph No. 6, Harvard Un. Press, Cambridge, 1946.
Z. Kopal: *The computation of elements of eclipsing binary systems.* Harvard Obs. Monograph No. 8, Harvard Un. Press, Cambridge, 1950.

SPHERICAL STARS

H. N. Russell: *Ap.J.*, **35**, 315, 1912.
H. N. Russell and H. Shapley: *Ap.J.*, **36**, 239, 1912.

OBLATE STARS

H. N. Russell and H. Shapley: *Ap.J.*, **36**, 60 and 397, 1912.
H. N. Russell: *Contr. Lick Obs.*, Series II, No. 23, 1948; *Ap.J.*, **108**, 388, 1948.

TABLES

E. Hetzer: *Dissertation*, Leipzig, 1931.
M. Wendt: *Dissertation*, Leipzig, 1931.
H. N. Russell and H. Shapley: *Ap.J.*, **36**, 243 and 390, 1912.
W. P. Zessewitsch: *Poulkovo Circ.*, No. 24, 41, 1938.
W. P. Zessewitsch: *Publ. de l'Institute Astronomique URSS*, No. 45, 1938; No. 50, 1940.
K. Ferrari: *Mitt. Un. Sternw. Wien*, **1**, No. 6, 422, 1939.
L. Plaut: *Groningen Publ.* No. 54, 1950.
J. E. Merrill: *Princeton Contr.* No. 23, 1950; No. 24, 1953.

DIFFERENTIAL CORRECTIONS

A. B. Wyse and G. E. Kron: *L.O.B.*, **19**, 17 and 28, 1939.
L. Binnendijk: *B.A.N.*, **9**, 182, 1941.
J. B. Irwin: *Ap.J.*, **106**, 380, 1947.

GRAVITATION EFFECT

H. von Zeipel: *M.N.*, **84**, 665, 1924.
S. Chandrasekhar: *M.N.*, **93**, 539, 1933.
H. N. Russell: *Ap.J.*, **90**, 653, 1939.
A. J. Wesselink: *Leiden Ann.*, **17**, part 3, 1941.

REFLECTION EFFECT

J. Stebbins: *Ap.J.*, **32**, 200, 1910.
H. N. Russell: *Ap.J.*, **36**, 67, 1912.
A. S. Eddington: *M.N.*, **86**, 320, 1926.
E. A. Milne: *M.N.*, **87**, 43, 1926.
Z. Kopal: *Ap.J.*, **89**, 323, 1939; *M.N.*, **114**, 101, 1954.

ECCENTRICITY

H. N. Russell: *Ap.J.*, **36**, 54, 1912.
F. Schlesinger and S. Udick: *Allegheny Obs. Publ.*, **2**, 155, 1912.

ROTATION OF THE LINE OF APSIDES

J. Stein: *Specola Astron. Vaticana*, **6**, 230, 1924.
H. N. Russell: *M.N.*, **88**, 641, 1928; *Ap.J.*, **90**, 641, 1939.
T. G. Cowling: *M.N.*, **98**, 734, 1938.
Z. Kopal: *M.N.*, **98**, 448, 1938.
T. E. Sterne: *M.N.*, **99**, 451, 1939.

EXTENDED ATMOSPHERES

F. E. Roach and F. B. Wood: *Ann. d'Astrophysique*, **15**, 21, 1952.

CATALOGUES OF ORBITAL ELEMENTS

C. Payne-Gaposchkin and S. Gaposchkin: *Variable stars*, Harvard Obs. Monograph No. 5, 67, 1938.

L. Plaut: *Groningen Publ.*, No. 54, 16, 1950; No. 55, 4, 1953.

Z. Kopal and M. B. Shapley: *Jodrell Bank Ann.*, **1**, No. 4, 141, 1956.

Index of Authors

Index of Subjects

www.ingramcontent.com/pod-product-compliance
Lightning Source LLC
Chambersburg PA
CBHW050657190326
41458CB00008B/2598
* 9 7 8 1 5 1 2 8 0 0 3 5 7 *